The Biology of
Agroecosystems

T0177816

THE BIOLOGY OF HABITATS SERIES

This attractive series of concise, affordable texts provides an integrated overview of the design, physiology, and ecology of the biota in a given habitat, set in the context of the physical environment. Each book describes practical aspects of working within the habitat, detailing the sorts of studies which are possible. Management and conservation issues are also included. The series is intended for naturalists, students studying biological or environmental science, those beginning independent research, and professional biologists embarking on research in a new habitat.

The Biology of Agroecosystems

Nicola P. Randall
Principal Lecturer, Crop and Environment Sciences,
Harper Adams University, UK

Barbara Smith
Associate Professor, The Centre for Agroecology, Water and Resilience,
Coventry University, UK

OXFORD
UNIVERSITY PRESS

OXFORD

UNIVERSITY PRESS

Great Clarendon Street, Oxford, OX2 6DP,
United Kingdom

Oxford University Press is a department of the University of Oxford.
It furthers the University's objective of excellence in research, scholarship,
and education by publishing worldwide. Oxford is a registered trade mark of
Oxford University Press in the UK and in certain other countries

© Nicola P. Randall & Barbara Smith 2020

The moral rights of the authors have been asserted

First Edition published in 2020
Impression: 1

All rights reserved. No part of this publication may be reproduced, stored in
a retrieval system, or transmitted, in any form or by any means, without the
prior permission in writing of Oxford University Press, or as expressly permitted
by law, by licence or under terms agreed with the appropriate reprographics
rights organization. Enquiries concerning reproduction outside the scope of the
above should be sent to the Rights Department, Oxford University Press, at the
address above

You must not circulate this work in any other form
and you must impose this same condition on any acquirer

Published in the United States of America by Oxford University Press
198 Madison Avenue, New York, NY 10016, United States of America

British Library Cataloguing in Publication Data
Data available

Library of Congress Control Number: 2019941136

ISBN 978-0-19-873752-0 (hbk.)
ISBN 978-0-19-873753-7 (pbk.)

DOI: 10.1093/oso/9780198737520.001.0001

Printed and bound by
CPI Group (UK) Ltd, Croydon, CR0 4YY

Links to third party websites are provided by Oxford in good faith and
for information only. Oxford disclaims any responsibility for the materials
contained in any third party website referenced in this work.

Acknowledgements

The authors would like to thank Simon Jeffery and Sue Everett for their contributions to the soil chapter, Matthew Burgess for his contributions to the Glossary, and Alfred Gathorne-Hardy, Ben Simmill, Bruce Schaffer, and Paul Lewis for their comments elsewhere. We would also like to thank Gavin Siriwardena for providing BTO farmland bird data.

Acknowledgements

Contents

1

Introduction

1.1 What are agroecosystems?

An ecosystem is a community of living organisms, the abiotic environment, and
the interactions that take place through flows of energy and nutrients, within a
given area. The subset of a natural ecosystem that is managed for agricultural production is known as the agroecosystem. The limits of an agroecosystem are not
easily defined, as the impact of agriculture is not limited to the land under production. Agroecosystems include cultivated areas (growing arable crops such as
grains, rice, and oilseeds, or horticultural crops such as vegetables), pasture and
leys that support livestock, permanent crops such as orchards or plantations, and
mixed farming systems.

Agriculture takes place in three of the world's four climatic zones: temperate, tropical, and arid (but excluding polar) at a variety of altitudes. Agroecosystems are
rarely homogenous and are usually a matrix of land managed for production and
peripheral non-farmed areas, such as scrub, grasslands, woodland patches, and
watercourses. Agroecosystems are highly variable, and their characteristics will
depend on the regional and local context; some will be comprised largely of monocultures whereas others will be comprised of highly diverse mixed cropping and
livestock systems. The amount of semi-natural habitat in the matrix will vary. All
of these factors are also determined by environmental variables such as local climate, geology, and altitude. As a result, both globally and regionally, we find significant variation within and between agroecosystems. Individual agricultural
sites range from small subsistence holdings to very large intensively managed
farms. For example, the average landholding in Asia is about 1 hectare (ha), and
these holdings are frequently even smaller in many regions (for example, holdings
in Bangladesh are estimated to average only 0.3 ha). This contrasts with North
America where the average size is nearly 120 ha (Asia and Pacific Commission on
Agricultural Statistics 2010). The one factor that ties agroecosystems together is
that all are managed, to a greater or lesser extent, for production (usually of food,
fibre, or fuel).

The Biology of Agroecosystems. Nicola P. Randall and Barbara Smith, Oxford University Press (2020).
© Nicola P. Randall and Barbara Smith 2020. DOI: 10.1093/oso/9780198737520.001.0001

1.2 Why study the biology of agroecosystems?

The Royal Society of Biology considers biology to 'encompass all areas of the science of life from molecules, through whole organisms to ecosystems' (https://www.rsb.org.uk/). The extent and impact of agroecosystems on the global environment, and their importance for sustaining the human population means that a good understanding of agroecosystems and their functions is essential.

Most global landscapes have been impacted by agriculture, and even unmanaged or semi-natural biotopes are surrounded by agriculture of some kind. Agricultural practices have global and ever-changing impacts on the biology of ecosystems. It is important to remember that agricultural environments are anthropogenic habitats both in their origin and ongoing maintenance, and so are inherently unstable, even in the case of low impact subsistence agriculture. The management decisions that we make will impact on both food production and the condition of the wider environment.

1.2.1 The extent and characteristics of agroecosystems

Humans have managed land for food production for at least 10,000 years. Nearly 40 per cent of the global land area is currently classified as agricultural, although this percentage can vary widely between continents and countries. The Falkland Islands, for example, had over 90 per cent of its land area classified as agricultural in 2015, whereas in Egypt, where most of the country is desert, this figure was less than 4 per cent (FAO 2018). There is also great regional disparity between the proportion of agricultural land that is cultivated as opposed to under permanent crops or pasture, and the proportion of land that is used for food production rather than for other products. Plate 1 shows the disparity between regions that primarily grew crops for human food consumption and those where alternative products dominated in the year 2000.

Agriculture has expanded in line with global populations, and in the twentieth century farming was transformed by the development of world markets and technology. In what has been called the 'Green Revolution', agricultural change was characterized by the development and distribution of hybridized varieties of cereal grains, the improvement and expansion of irrigation, the introduction and distribution of synthetic fertilizers and pesticides, and the modernization of management techniques including increased mechanization.

Whilst achieving increased food production, this expansion and intensification of agriculture has resulted in environmental impacts, such as increased levels of diffuse pollution within water catchments and reduced heterogeneity within the landscape. Declines in farmland wildlife, have been frequently recorded (Benton et al. 2003), (including birds (Donald et al. 2001), invertebrates (Molina et al. 2014), and wildflowers (Sutcliffe and Kay 2000). Natural biotopes are increasingly fragmented as agricultural activities expand around them, and this trend is likely to continue.

Agroecosystems will continue to change in the future. Global human populations rose from 2.5 billion in 1950 to 7.6 billion in 2017 (United Nations, Department of Economic and Social Affairs, Population Division 2017), and this, coupled with changing dietary patterns, has led to pressures for further increases in food production. Scientists predict that global crop demand will have at least doubled between 2005 and 2050 (Tilman et al. 2011). Boosting productivity through new methods and technologies is one way to meet this challenge, but is unlikely to meet these increasing demands alone (Ray et al. 2013), and so further intensification of existing agricultural systems is likely to be combined with increased conversion of land for agriculture. The most intensive agriculture tends to be on arable land. Around 11 per cent of global land surface is currently either in arable land or under permanent crops. The FAO predicts that this only represents just over one-third of land that has crop production potential, although how much of this is likely to be brought under future cultivation is disputed (Bruinsma 2017).

This evolution of agriculture helps to ensure that increasing food production requirements are met, but the natural biological components that facilitate ecosystem services, such as clean water, pollination, and bio-control, are also essential to support this increase (Zavaleta et al. 2010; Blitzer et al. 2016). In order to understand how to best deliver sustainable agroecosystems for the future, an understanding of the relationships between land use, ecosystem services, and the related organisms that deliver them is necessary (Bommarco et al. 2012).

1.2.2 Studying the biology of agroecosystems

The study of agroecosystem biology necessarily includes a diverse range of topics. Studies of agroecosystems may focus on the crops and livestock farmed, or the physical changes facilitated, such as management of water tables through drainage, damming, or irrigation, or from reprofiling of the land itself such as in regions where terracing of hillsides is common. Agroecosystems and their components can be studied at a variety of scales, from specific organisms to the multiple interactions within a landscape and beyond.

In addition, agricultural activities may impact on non-agricultural ecosystems. For example, Tilman et al. (2001) predicted that agricultural growth would lead to a more than 2.4-fold increase in eutrophication of terrestrial fresh waters and of coastal marine systems due to nitrogen and phosphorus increases.

To complicate matters, food production has implications beyond biology. Societal, political, economic, and technological factors are as important in the consideration of sustainable agriculture as biodiversity and ecosystem services. These external drivers influence agricultural biology at a local and global scale, which in turn impacts on ecosystem functions. Likewise, it is the agroecosystems and the biological relationships that take place within them that often drive the social, economic, and political decisions that are made, from individual farmers to international bodies.

1.3 The purpose and organization of the book

This book focuses on the biological aspects of terrestrial agriculture, but will consider these in the wider agroecological context. The book provides an introduction to the biological and ecological attributes of agroecosystems, and the biological impacts of agriculture. It is designed to aid the reader's understanding of agroecosystems, the biological pressures that they face, and the issues that must be considered in their management.

The book introduces different types of farming systems and their associated biology. It illustrates some of the specific physiological and ecological issues associated with agriculture, together with potential management options for reconciling production of food and other commodities (fibre, medicines, energy, etc.) with conservation of natural resources and biodiversity.

Following on from this introductory chapter, Chapter 2 gives an overview of the historical context of modern agriculture and the key characteristics of agroecosystems. Chapter 3 examines general biodiversity within and between different types of farming systems, and outlines the relationship between agriculture and patterns of biodiversity in space and time, from the genetic to the landscape level. Chapter 4 builds on this by considering biotic interactions and the role of functional groups within agroecosystems. Chapters 5 and 6 address the specific key ecosystem components, soil and water, and consider how agricultural practices affect and are affected by these key components. Chapter 7 discusses the impact of the globalization of agriculture on the biology of agroecosystems. Chapters 8 and 9 discuss the sustainable management of agroecosystems and look to the future challenges for agroecosystem biology.

Case studies from around the world are used throughout the book to illustrate the wide variety of global farming systems and the biological issues and solutions associated with them.

2
Agricultural Environments

2.1 Introduction

This chapter outlines the origin and development of global agriculture. Agricultural systems around the world are extremely diverse, but all involve the management of land for the production of food and other commodities.

Agriculture is defined as the science of farming, which is itself defined as the activity of growing crops or raising livestock (Oxford English Dictionary 2004). Food is the most widely produced agricultural product, but diverse agricultural commodities are produced throughout the world. Non-food commodities include fibre for clothing, biomass crops for energy, wood for construction, ingredients for medicines and beauty products, and flowers for décor.

The global extent of agricultural land accounts for 37.6 per cent of the land area on Earth (FAO 2013; Roser 2016); this can be subdivided into three broad categories: arable land comprising 28 per cent, permanent crops 3 per cent, and permanent meadows and pastures 69 per cent. The number of hectares dedicated to agriculture has changed very little since 2009 and is not predicted to rise appreciably in the foreseeable future (Roser and Ritchie 2017). Arable production, however, has risen steeply; by 2012, just 32 per cent of the land area could produce the equivalent amount of crops to all of that needed in 1961 (Ausubel et al. 2013). A combination of agricultural technologies and changing tastes means that there has been downward pressure on croplands (Ausubel et al. 2013), and it is expected that further increases in crop production are also likely, frequently due to intensification of farming rather than range expansion.

2.2 The origins and historical development of agriculture

Agriculture requires the cultivation, domestication, and management of plants and animals. Before humans were farmers, they were hunter-gatherers. Some 15,000 years ago, humans foraged for fruits, seeds, and grains, supplementing these with protein from hunting and fishing. They led nomadic lifestyles, moving

The Biology of Agroecosystems. Nicola P. Randall and Barbara Smith, Oxford University Press (2020).
© Nicola P. Randall and Barbara Smith 2020. DOI: 10.1093/oso/9780198737520.001.0001

seasonally depending on food availability. The development of agriculture has been pieced together from archaeological and paleontological records, supplemented by anthropology and molecular studies. From these, we know that wild grains were collected and eaten from at least 20,000 BC, and it is generally agreed that agriculture spontaneously and independently arose in several regions somewhere between 15,000 and 10,000 BC, although estimates vary. Vasey (1992) outlines evidence of scattered agricultural practices that range from 8,500 years BC in Southwest Asia, Peru, and Mexico, to 12,000 BC in Taiwan based on evidence of regular forest burning, and as far back as 14,000 BC in the Nile valley based on the tools that have been found. Emmer wheat, einkorn wheat, hulled barley, peas, lentils, bitter vetch, chick peas, and flax (the eight Neolithic founder crops) were cultivated in the Eastern Mediterranean from around 9,500 BC.

At least 11 regions of the Old and New World have been identified as centres of domestication (Larson et al. 2014). The earliest examples are from Southwest Asia, where there is evidence of domestication of wheat, barley, lentils, peas, sheep, goats, pigs, and cattle. In Mesoamerica, squash was domesticated in the early Holocene (Larson et al. 2014). Domestication changes crop species traits, towards palatability and ease of management. There is evidence in the botanical remains from human teeth that suggest the loss of fruit bitterness in *Cucurbita moschata* (a squash species) had taken place by 11,500 BC (Larson et al. 2014). The selection for desirable traits was an ongoing process of both conscious and unconscious selection (Darwin 1868). For example, it has been suggested that the glutinous quality of rice was selected for its taste, but that the non-shattering of cereal seeds is due to the habit of harvesting of cereals by sickle, which favoured plants with seeds that do not fall from the stalk (Larson et al. 2014). In animals, domesticated traits include increased docility, altered reproduction patterns, and changes in body proportions (Hammer 1984). Pigs, native to Eurasia and Africa, were domesticated in Mesopotamia *circa* 11,000 BC; cattle are thought to have been domesticated from a type of wild cattle, aurochs, in two separate events in the Indian subcontinent and in Western Asia around 6,000 to 8,000 BC; and in South America people domesticated llamas, alpacas, and guinea pigs.

The widespread adoption of agriculture came with the Neolithic (approximately 1,000–2,000 BC). The first agricultural civilization was Sumerian, in southern Mesopotamia (a region in Western Asia). The Sumerians were living in villages by 8,000 BC and cities by 5,500 BC, having developed a system of irrigation that allowed them to grow grain at scale using the waters from the Tigris and the Euphrates, despite low rainfall in the area (Wilkinson 2013). However it was the Egyptians who practised agriculture on a large scale between 10,000 and 4,000 BC, allowing them to build an Empire on their agricultural wealth (Janick 2000). Subsequently, the Greeks and the Romans drew on the expertise of these early agrarian societies (Mazoyer and Roudart 2007).

The development of the plough was key to the growth of agriculture, but its origin is uncertain. The earliest archaeological evidence for a ploughed field is from the

Indus valley *circa* 2,800 BC (Lal 2003) and there are pictograms of ploughs from Sumaria in 3,000 BC. These early implements were largely wooden. When iron came into widespread use in 500 BC, mechanization, particularly the use of the plough, facilitated a profound change in both society and the environment by enabling rapid deforestation and cultivation of the land. Early Roman agriculture was based on small, hand-cultivated landholdings of 1–5 acres (0.4–2 ha). With the advent of ox and plough, a single individual was able to prepare a much larger plot. Increasingly, Roman agriculture specialized, particularly in vines and olives, and the many small farmers made way for fewer, large plantation owners. The small farmers, no longer occupied by agriculture and pushed into the cities, were then available to be incorporated into the armies that would be used to forge the Roman Empire (Montgomery 2007).

At the same time as the Romans, great cities were supported by agriculture in Meso- and South America. The Aztec, Maya, and Inca civilizations developed terracing, raised bed technologies, and irrigation systems, which supported population growth. Agriculture was based on corn in Mesoamerica and potato in South America, both crops which are still hugely important in modern agriculture and currently support a large proportion of the world's population (Mazoyer and Roudart 2007).

A suite of major advances in agriculture occurred during the 'Arab agricultural revolution', which refers to the development of agriculture in the medieval Arab empire and its diffusion from Europe to the borders of the Far East. The Arab trade routes enabled the exchange of novel crops. This was an era of scholastic enterprise. Islamic scholars, particularly botanists, experimented, taught widely, and published scholarly works such as the twelfth-century *Kitab al Filaha* 'Treatise on Agriculture', and offered practical advice, such as the Calendar of Cordoba, an agricultural calendar, which they translated and distributed. During this time, irrigation technologies were developed, understanding of soil fertility management advanced, and the introduction of high-value crops and animals proliferated, supported by the distribution of free seeds, advice, and education. Relatively liberal landownership laws were also introduced. Anyone was allowed to own, sell, and inherit land. In contrast to the Roman model where serfs were tied to (and sold with) land, those who directly worked the land had rights to a share of the produce. These circumstances, combined with central government policy to exploit undeveloped lands, resulted in advancements in agriculture (Idrisi 2005).

In northern Europe, the eleventh and twelfth centuries saw an expansion of the monastic system when the monasteries became centres of expertise for agriculture and forestry. The church had large landholdings and in conjunction with the manorial system, under which large landholders farmed using indentured serfs, landowners had control over large areas of land. New crops were imported via Arab traders. In 900 AD the introduction of the 'heavy plough' enabled deep ploughing of heavy, more clay-based soils which greatly increased productivity.

Land was drained and marginal lands were brought into production. The adoption of a three field rotation system (rather than two) added nitrogen-fixing legumes to the rotation which resulted in the improvement of soil fertility. Windmills and watermill technology improved, leading to increased efficiency of processing both food and fibre. Crop yields stabilized in the thirteenth century, with the next significant shift in agriculture coinciding with the 'discovery' of the Americas and the exchange of crops between the Old and New Worlds in the fifteenth century. Horses, cattle, sheep, and goats were introduced into the New World, which had few livestock species, and European traders introduced cassava and maize into Africa, replacing traditional African foods (Wambugu and Wafula 2000). In Europe, the introduction of the potato generated great change. Potatoes are calorie dense and the impact of their introduction was to double Europe's food supply (in terms of calories) and reduce famine (Mann 2011). A secondary impact was the necessity to adopt the intensive fertilization required by potatoes, which was helped by the introduction of guano from Peru. This set the direction for an agricultural system where fertility needs were met by the addition of imported nutrients.

The advent of what we might consider 'modern farming' arrived in the seventeenth to nineteenth centuries with the agricultural revolution, centred in Europe and largely in the UK, sometimes called the British Agricultural Revolution, which was a process of increasing mechanization (Overton 1996). There was an explosion of new technologies and changes to farming. In Britain, this period of change was underpinned by the 'Enclosure Acts' which enclosed common lands and passed them into private hands and removed the existing manorial system whereby tenant farmers were only able to farm small strips. This transfer of power and ownership gave a newfound confidence to landowners, who invested heavily in land that gave them the whole return (Overton 1996). However, it was a process that took place over many years; there were eleven Enclosure Acts, the first in 1773 and the last in 1859. In total, over 5,200 individual Enclosure Acts were passed, covering 2.75 million ha (House of Commons 1919). There were several key innovations in this period that led to great increases in productivity and efficiency:

- **The development of the seed drill**. Before the advent of the seed drill, seeds were broadcast by hand. The seed drill was imported into England from China, and developed by Jethro Tull in 1701; it distributed seeds evenly and at an optimum depth, and continued to be developed and improved over the next hundred years. It improved germination, allowed planting in rows to make husbandry easier, and, by sowing beneath the soil, minimized seed predation.
- **The development of mechanical threshers**. Threshing, the process of separating seed from stalks and husks after harvest, was a laborious and time-consuming process. In the 1780s, Andrew Meilkle, a Scottish engineer, developed a mechanical horse-powered alternative to traditional hand threshing. The model was improved and developed incrementally in the UK, Australia, and America and was the precursor of the combine harvester. It greatly reduced the labour associated with harvest, but at the expense of many rural jobs.

- **The four crop rotation system.** This system was developed in the Netherlands in the seventeenth century but popularized by Charles 'Turnip' Townsend in the eighteenth century in Britain. The system comprised a rotation of four crops: barley, wheat, turnip (as fodder), and clover (as grazing), which enabled animals to be raised all year round.
- **Selective breeding of livestock.** Robert Bakewell and Thomas Coke made great innovations in the selective breeding of livestock and are particularly well known for the Leicester Longwool sheep which gained weight quickly and survived well on turnips. It was the first purebred sheep to be imported into Australia. Although possibly the most famous example, there was extensive experimentation with selective breeding in this period (Stanley 1995).

These innovations were exported and further developed all over the world, changing the face of agriculture. At the end of this period (1910), Fritz Haber and Carl Bosch perfected their method to convert atmospheric nitrogen to ammonia, a process that is the basis of contemporary artificial fertilizers. The process is now used to produce 450 million tonnes of nitrogen fertilizer a year and has fuelled our ability to feed a growing population—so much so that it has been called the 'detonator of the population explosion' (Smil 1999). Erisman et al. (2008) calculated that by the early 2000s nearly 50 per cent of the world's population was fed by use of the Haber–Bosch process, and without it the world's population would be less than four billion (rather than the nearly seven billion that it was at the time).

2.3 Contemporary agriculture

Post World War II, there was further intensification of agriculture, notably through chemical inputs such as fertilizers and pesticides, selective breeding of crops, and the development of high-yielding hybrids, combined with large-scale motorization and mechanization, particularly in the more developed countries. A variation of this, but without the large-scale motorization and mechanization, occurred in developing countries and was termed the 'Green Revolution' (Mazoya and Roudart 2007).

Currently, agriculture covers nearly 40 per cent of the global land area. Agricultural systems are diverse and encompass a variety of production systems, under varying levels of intensification. They can include a range of food and non-food systems (e.g. cereals, meat, vegetables, fibre, medicines, and energy), but all involve the management of land for the production of food and other commodities. In addition to the land that is managed to support agricultural production, many agroecosystems contain semi-natural areas or land that is managed for other purposes, which introduces another layer of complexity and variation into these ecosystems. Figure 2.1 illustrates this complexity.

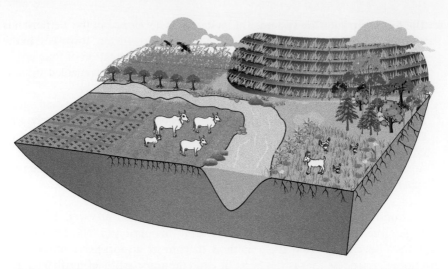

Figure 2.1 An example of a mixed agroecosystem, illustrating the complexity within and between farming systems and the influence of local environments. Features include (clockwise from bottom left): arable crops, wooded strip, rough grassland, terracing, woodland/agroforestry, mixed live-stock and cropping, intensive livestock. A watercourse runs through the agroecosystem.

2.3.1 Spatial patterns of agriculture

Agriculture is practised on all of the Earth's continents except Antarctica. The extent and configuration of what can be considered agricultural land varies greatly within and between regions, as does the intensity of agriculture. The Falkland Islands, for example, which had over 90 per cent of its land area classified as agricultural in 2015 (FAO, 2018), consists mainly of extensive rangeland pasture. This contrasts with Europe, where less than half of the land is agricultural, with nearly 60 per cent of this classed as arable and only 34 per cent permanent pasture (European Union 2018). Dietrich et al. (2012) used a ratio measurement to standardize extraneous factors, such as soil and climate, to establish levels of agricultural intensity. They found that Africa, Asia, and parts of Russia were predominantly low intensity, whilst the Eastern United States, Western Europe, and parts of China had a general trend towards high intensity.

Some of the major global agricultural systems are shown in Plate 2.

Agricultural systems are still expanding and becoming more intensive. The world's cultivated area grew 12 per cent in the 50 years up to 2010, and in the same period irrigated systems doubled (FAO 2011).

In some regions, agriculture is increasingly spreading into marginal lands (those that have limitations for agricultural production, and can only support low production). In 2013, marginal lands accounted for about 36 per cent of global agricultural land (1.3 billion ha), and supported roughly one-third of the world's population, although the use of marginal lands constantly changes along with socio-economic conditions, such as market demand, land rents, and profits, which may push and pull marginal land in and out of production (Post et al. 2013).

People may be forced to farm marginal lands, but at times when there is less pressure to produce food on them they can be important for second-generation biofuels such as woody crops (Tilman et al. 2009), or set-aside for biodiversity.

While agriculture expands in some areas of the world, in others, areas that were previously cultivated are being abandoned. The principal drivers for land abandonment are: rural–urban migration stimulated either by the pull of economic opportunities in cities or the push of armed conflicts, political instability, or large-scale natural catastrophes; ecological drivers such as land degradation of land due to mismanagement; loss of local production incentives due to increased yields on productive lands, conservation policies, or as a result of increased imports; reduced profitability of local farming; technology, industrialization, land tenure systems, or access to markets (both physical and economic); and lack of economic and demographic viability—this is particularly associated with small and family farms. In Europe, poor environmental suitability for agriculture, low farm stability and viability, and the impact of subsidies from the Common Agriculture Policy (CAP) have been important factors. Low income, low investments on the farm, and an aging population of farmers are also implicated (Leal Filho et al. 2017).

2.3.2 Agricultural classifications

Classification of agricultural systems may appear simple at first glance. Systems are arable (crop based), pastoral (with livestock), or mixed. They can be intensive (with high inputs of labour and/or capital) or extensive. But there are many factors other than intensity and production system that are taken into consideration when classifying agriculture. A number of systems have classified cultivated areas of the earth into agroclimatic zones, based on major climate and growing period which determine the suitability for a certain range of crops and cultivars (Van Wart et al. 2013). Zones range from warm tropics through to arctic. These have been further subdivided into agroecological zones, where the zones are refined using altitude, soil variables, and crop-specific constraints (IIASA/FAO 2012). Other classifications consider whether the system is subsistence (where most of the product is used by the farmer and his family) or commercial, sedentary, or nomadic. Environmental conditions and whether or not a system is rainfed or irrigated can also be used to differentiate between systems. Because of this complexity, there have been numerous attempts made to classify agricultural systems, and no one system is used over all others. Robinson et al. (2011) summarize some of the different classification systems in use (Table 2.1).

2.4 Characteristics of agricultural systems

Even within classifications, the variations within and between farm systems are complex. For example, in pastoral systems, the livestock can be as variable as the systems themselves, and could include sheep and cattle in North America, yaks, camels, goats, and horses in Central Asia, and reindeer in circumpolar regions. Box 2.1 gives some examples of contrasting farming systems.

Table 2.1 Comparison of some existing farming system classifications and methods[a]

Classification	How are crops dealt with?	How are livestock dealt with?	How many categories?	Pros and cons, and can it be mapped?
Ruthenberg (1980)	1. Degree of cultivation (rotation) 2. Forest, bush, savanna, grass 3. Crop type 4. Irrigated vs rainfed	Degree of movement/permanence	8 major	Categories too broad and incomplete
Grigg (1972) after Whittlesey (1936)	1. Crop type 2. Commercialization 3. Location/agroecology	Degree of movement/permanence	9 major	System incomplete and somewhat selective
Dixon et al. (2001)	1. Crop type 2. Commercialization 3. Location/agroecology	Degree of movement/permanence	8 major, 72 globally (type by region)	Derivation not explicit, may be difficult to map using existing global datasets
Seré et al. (1996)	1. Are there crops or not? 2. Rainfed vs irrigated 3. Agroecology	1. Landless or rangeland based 2. Agroecology	11 major	Livestock-based, so no categorization of crop systems Can be mapped approximately using appropriate proxies
Explicit AEZ methods, e.g. Fischer et al. (2002)	Match land suitability to crop requirements for given inputs and technology	Not dealt with, though probably could be included	As required	Easily mapped. Assesses what may be rather than what actually is
Statistical classifications, e.g. Wint et al. (1997)	Cluster spatial units based on crop densities and intensities	Cluster spatial units based on livestock densities	As required	Easily mapped. Arbitrary, data sensitive, and non-replicable

[a] Numbers in columns showing how crops and livestock are dealt with broadly indicate stages in classification.

AEZ, Agroecological zone.

Source: Robinson et al. 2011.

Box 2.1 Examples of some contrasting farming systems from around the world

Nomadic reindeer farming

In subarctic regions across Europe, Asia, and North America, semi-domesticated reindeer are used for food, hides, milk, and transport. What probably began as part of a hunter-gatherer lifestyle developed into reindeer pastoralism. Herders and their reindeer migrate from winter grazing to rich northern, upland, or coastal pastures in summer months. Reindeer are ruminants and are able to gain nutrients from lichens through foregut fermentation when other food is scarce in the winter months. Many modern herders are part of co-operatives of some sort, and herds in the tundra regions may number in the thousands and travel over 1,000 km per year. Mechanization is increasingly used, including helicopters and snowmobiles, and this means that some herds are not always accompanied throughout the year (Figure 2.2).

Intensive dairy farms in China

Between 1990 and 2017, China grew from being the 24th largest dairy producer in the world to the 4th largest, and in 2014 had nearly 1,500 farms with herds of over 1,000 head (although the majority of farms are still owned by peasant farmers with only a few animals). The dairy cows are fed on a mix of grains and fodder and

Figure 2.2 Nomadic reindeer farming. Photo by Heather Sunderland, reindeer man feeding the reindeer in Lapland. Used under the Creative Commons Attribution License 2.0. No changes have been made.

Continued

Box 2.1 *(Continued)*

milked three times a day on a rotary system. These high-tech systems utilize specialist breeding and feeding techniques, which allow the cows to be housed indoors throughout the year, and boost the outputs. Artificial insemination is used to maintain the herds.

The Mudanjiang Mega dairy farm in the northeast of China has 100,000 cows producing 800 million litres of milk a year. The cows are kept in large barns in which a combination of breeding and specialized diets has increased milk production by 30 per cent. However, the farm has been criticized for releasing a range of pollutants into the local environment including pesticides, hormone and antibiotic treatments, run-off of manure into watercourses, and high methane emissions from the cows themselves. Researchers have estimated that in China in 2013 the dairy industry produced 225 million tons (MT) of manure and was responsible for the release of NH_3_N ammonia (0.2 MT), nitrogen (0.5 MT), and phosphorus (0.04 MT) into the environment (Qian et al. 2018).

Rice in paddy fields

In Asia, much rice is grown in paddy fields (partially flooded arable fields) (Figure 2.3). While China produces the most rice of any country globally, India has the largest area of rice production in the world, and in many parts of India the rice is cultivated twice, or even three times, a year. The establishment of rice in fields varies according to variety and location, but typically fields will be puddled before cultivation. This involves repeated cultivation in waterlogged conditions, specifically designed to break down the soil structure so that individual soil particles fit

Figure 2.3 Rice in paddy fields. Photo by Melanie_ko, paddy fields. Used under the Creative Commons Attribution License 2.0. No changes have been made.

closely together to create a pan that keeps the water from draining freely away. This is in contrast to nearly every other arable crop, where soil structure is maintained or enhanced wherever possible. Rice plants are grown in nurseries (normally at the side of a field) and then transplanted into the main field, usually by hand but increasingly with machines.

A key purpose of flooding is weed control, as rice plants (in contrast to most other plants) have aerenchyma, special tubes in their stem to take air (including oxygen) down to their roots.

Chakra systems in the Amazon Basin

Chakra systems are agroforestry systems traditionally practised in the Amazon Basin. Chakras are patches of modified forest lightly managed for food production. There is high diversity of plants, and up to 50 species of perennials and annuals are grown together in a single plot, including cocoa, guava, guaba (*Inga* spp.), citrus, coffee, plantains, pineapple, cassava, sweet potato, corn, peanuts, several varieties of beans, palms including peach palm (*Bactris guisapeas*), and a range of medicinal plants and timber trees. They are owned collectively, with individuals being allocated a plot. The size of plot depends on land available and the number of people in the collective but can be up to 50 ha (Krause and Ness 2017).

Hill farming in the UK and Ireland

Hill farming is an extensive farming system, commonly raising sheep (although with some cattle), which are grazed at an intensity of about 2 ha per head (Figure 2.4). The land is poor and usually part of land categorized as Less Favoured Area (LFA), which is land that is characterized by poor-climate soils and terrain. Usually the vegetation comprises nutrient-poor grasslands and heath. Low nutrient status means that the animals tend to be lean and it is very common to move animals to the richer lowlands to fatten them for market. With low incomes, these marginal farms are supported by subsidy through central government. It is considered valuable for the way of life it preserves and because the farming system supports a diversity of flora and fauna in comparison with more intensive agricultural habitats.

New Zealand mixed farming

Mixed farming is practised in New Zealand to varying degrees of intensity (Figure 2.5). The key characteristic is that mixed farms comprise both crops and livestock, or mixed livestock. They may be simple farms with low crop diversity, for example arable cropping combined with dairy farming, or they can be highly diverse with a range of animals and a range of crops, which could potentially include cereals, vegetables, and orchards. Mixed farms are not necessarily organic and farmers will use the normal range of pesticides and fertilizers. However, organic farms are frequently mixed farms due to the benefits of having animals on site for nutrient provision.

Continued

Box 2.1 (*Continued*)

Figure 2.4 Sheep grazing, Republic of Ireland. Photo: Nicola Randall

Figure 2.5 New Zealand mixed farm. Photo by Bernard Spragg

2.4.1 Temporal patterns in agricultural systems

Most agroecosystems are subject to more short-term temporal change than natural ecosystems, particularly in cropping systems, but the speed of change varies from system to system. Some systems undergo rapid change but always within an agricultural regime. For example, in India, rice can be harvested once, twice, or three times a year, depending on the region, variety, and growing season. Other systems may shift from intense intervention to very low intervention. This is again illustrated in India, where small areas of land are carved out of naturally vegetated areas and cultivated for a single season, followed by several years of regeneration. This type of agriculture, known as shifting cultivation, also takes place in the Philippines, but here, rather than cropping for a single season, the land is usually cropped for a few years before being left to regenerate. This again contrasts with permanently established areas, such as the 168,000-ha Los Alcornocales cork oak forest in southern Spain, where lightly managed cork oak trees are harvested on a 7-year cycle. In many regions, agroecosystems may consist of areas with multiple different temporal patterns. In the UK, for example, a rotational system of cereals and a brassica is common, but there are also grasslands and permanent pastures which change very little from season to season or year to year. Regional variation is impacted by both between- and within-farm variability.

2.4.2 Pest management in agroecosystems

The management of pest species is a key challenge for agriculture. A pest is a species that competes with human needs by harming humans, domesticated animals, or cultivated crops (Hajek 2004). In agricultural systems this harm usually relates to the latter two, and thereby impacts on productivity of the animals or plants being farmed. Agricultural pests are hugely varied, consisting of vertebrate and invertebrate animal species, plants (weeds), fungal and bacterial pathogens, and viruses. Pests are common in agricultural systems, encouraged by the simplified, monocultural systems that do not support the diverse communities of pest predators and pathogens that might otherwise control them, and which provide an unnaturally high concentration of resources which in turn can support large pest populations.

There are multiple methods controlling pests in agricultural systems that can be roughly divided into three pathways: (1) husbandry, (2) chemical intervention, and (3) ecological management.

2.4.2.1 Husbandry

Husbandry comprises a range of low-input management practices which can help control and discourage pests during agricultural production. Manual or mechanical weeding is used in organic farming systems to lower the weed burden which competes with crops for growth. The invention of the seed drill, by creating straight rows of standard widths, greatly enhanced the ability of farmers to mechanize the process with mechanical weeders. Weeds can also be suppressed with the use of organic or plastic mulches. The disadvantage of these methods is that they are still relatively labour intensive and, although it is possible to reduce weeds between the rows, they do not address weeds within rows around the crop plants. Living mulches, i.e. plants that are intentionally sown between crops to provide ground cover, can also be used to control weeds, and can be incorporated into the soil after harvest to deliver other benefits such as nutrients and organic matter. In addition, plants can be sown or harvested at a time to avoid specific pests, and the use of rotations can help reduce build-up of pests and diseases in soils by breaking their lifecycles. In animal husbandry, keeping animals at reduced densities and in clean conditions can reduce pest species. This is especially relevant when animals are kept in enclosed conditions and there is a high risk of pest or pathogen transfer.

2.4.2.2 Chemical interventions

A principal focus in modern agriculture is utilizing technologies to combat pest problems, including chemical interventions, particularly herbicides, rodenticides, insecticides, molluscicides, and fungicides. The majority of these are synthetic chemicals, although increasingly there is research into biopesticides, based on natural plant extracts that are perceived to have less impact on the wider ecosystem.

Globally, herbicides are the most commonly used crop protection compounds (Burke and Bell 2014). Herbicides aim to control unwanted plants within the agroecosystem and the mark of success for a herbicide is that it kills the 'weed' but not the crop.

Herbicides may be **selective**, such as the group of herbicides that are graminicides which only kill grasses, or those that only kill broadleaf species, such as 2,4-dichlorophenoxyacetic acid (usually known as 2,4-D). Selective herbicides work by differential absorption, or may exploit morphological or phenotypical differences between plants. Alternatively they may be **non-selective**, such as the broad-spectrum herbicide glyphosate, which controls a wide range of plants. Both glyphosate and 2,4-D are systemic herbicides, which work by entering the plant system and then moving to the site of action; systemic herbicides are effective for perennial weeds as the herbicide reaches the root system. However, they are relatively slow, taking up to 3 weeks to be effective. Contact herbicides work through foliar application, and are more rapid but require good coverage to be effective.

Herbicides can be targeted at different stages of the lifecycle; herbicides applied pre-emergence work by preventing seed germination; applied post-germination, they act on the growing plants. Since the introduction of 2,4-D in the mid twentieth century, herbicides have become more targeted towards specific genera of plants, and comprise cocktails of more than one active ingredient, tailored for crop-specific requirements and becoming increasingly selective. The major impact of herbicides has been to reduce labour and increase crop production. Their use has facilitated reduced tillage and no-till systems and enabled the abandonment of rotational systems as high-value crops can be grown in the same field year after year with no significant build-up of pernicious weeds (Burke and Bell 2014).

The use of compounds and extracts to control pests is not a new phenomenon; Homer reported sulphur being used as a fumigant as long ago as 1,000 BC. In 1867 an arsenic-based compound to control the Colorado potato beetle in the USA became the first widely used pesticide; but it was only just after World War I that a combination of improved chemical synthesis and large-scale production meant that the use of chemical pesticides soared (Hajek 2004). Insecticides are hugely important in agriculture to control invertebrate pests. Insecticides increase productivity through protecting crop yields and by controlling the insect vectors of disease (Aktar et al. 2009).

Like herbicides, insecticides can be either systemic (where ingestion of a treated plant infects the target insect) or contact (where the insect needs to come into direct contact with the product). One of the most widely used group of insecticides in the late twentieth and early twenty-first century has been neonicotinoids, which impact the invertebrate nervous system by encouraging the nerve fibres to fire continually until they fail, causing death (Cressey 2017). The first neonicotinoid to be released was in the 1990s, imidacloprid. Neonicotinoids are systemic pesticides often applied as a seed dressing, thereby reducing the problems associated with spray drift.

In livestock systems, one of the key disease management tools is the use of antibiotics, particularly in more intensive systems, where large numbers of animals may be grouped together in high densities. Antibiotics are usually applied as a feed (or water) additive, or as a prophylaxis. In addition to controlling or preventing disease, antibiotics are sometimes used to enhance growth (Bonfoh et al. 2010). Overall, global antibiotic use in livestock systems is difficult to estimate, but the US Food and Drug Administration reported that nearly 11 million kg of antimicrobial drugs were sold and distributed for use in food-producing animals within the USA in 2017 (FDA 2018).

2.4.2.3 Ecological management

It is possible to enhance natural control of pests by supporting their predators, parasites, and pathogens. This often takes place through management of the habitats that the biological control organisms rely upon. Ecological management for pest control is explored further in Chapter 4.

2.4.3 Plant and animal selection

Another characteristic common to most agricultural systems is the selective breeding and introduction of many of the farmed species.

Individuals in a natural population of a given species vary due to differences in their genes, or, more specifically, their alleles (i.e. the variant form of a specific gene expressed in an individual, such as colour variation). The relative frequency of alleles varies in different populations, and changes over time, partly due to natural mutations, but also in response to environmental conditions. Individuals with well-adapted characteristics (or traits) to a particular environment will be likely to survive and pass on these characteristics to their offspring, thereby increasing their frequency in the population. For example, the allele for long hair may be more prevalent in a small mammal population in a cool environment, when compared with a population of the same species in a warmer environment. Thus 'natural selection', first coined by Charles Darwin, is a change in the relative alleles that express a particular trait in a species. This means that all species naturally evolve over time in response to external drivers.

Most agricultural species have been subject to human intervention to encourage traits that make them more desirable for agricultural purposes. This artificial selection or 'selective breeding' process has been taking place over thousands of years to encourage desirable agronomic traits, such as size, yield, and quality, in both plants and animals. The traits that are considered desirable may change over time as farmers and consumers change their priorities. For example, in the dairy industry, cattle have traditionally been bred to maximize yield. However, more recently, the focus has changed towards the reduction of input costs (health, fertility, feed efficiency) (Sölkner et al. 2007).

In addition to the artificial selection of desirable traits in specific species, their founder species often occur many miles or even continents away from the agroecosystem in which they are farmed. For example, potatoes originated from South America, cattle and sheep from Europe and Asia, and all of these are now bred and farmed across the globe. The use of ruminants in agriculture (cows and sheep, for example) is particularly widespread; ruminants have a specialized feeding system containing multiple microbes, which enables them to break down plant material that is indigestible to humans. This allows them to convert low-nutrient organic materials into high-protein food products (meat, milk) as well as providing other products (fibres, manure, etc.). Consequently, ruminants can be farmed on land that is otherwise unsuitable for food crop production (Huws et al. 2018). Global populations of cattle were estimated to be 1.43 billion, and sheep and goats 1.87 billion in 2014, with a billion people dependent upon them for food security (Robinson et al. 2014).

2.4.4 Nutrients on farmland

Another common characteristic of many agroecosystems is the artificial manipulation of nutrients in the system. As plants and livestock are removed from agroecosystems to be used for food or fibre, nutrients are lost from the system along with the products, and this influences the management of the systems. This management may involve moving between different land areas, as in nomadic or shifting cultivation systems. It may involve manipulating planting and livestock management in order to reduce or replenish nutrients in a semi-natural way, such as planting nitrogen-fixing plants within the system. Alternatively, nutrients may be introduced into the system from elsewhere.

The nutrients that are essential for plant and animal growth are listed in Table 2.2. The elements that are essential for plant growth are divided into macro- and micronutrients, with micronutrients only needed in small amounts. The three

Table 2.2 Nutrients that are essential for plant and animal growth

Element	Essentiality		Critical leaf concentrations (mg/gram of DM)	
	Plant	Animal	Sufficiency	Toxicity
Nitrogen (N)	Yes	Yes	15–40	
Potassium (K)	Yes	Yes	5–40	> 50
Phosphorus (P)	Yes	Yes	2–5	> 10
Calcium (Ca)	Yes	Yes	0.5–10	> 100
Magnesium (Mg)	Yes	Yes	1.5–3.5	> 15
Sulphur (S)	Yes	Yes	1.0–5.0	
Chlorine (Cl)	Yes	Yes	0.1–6.0	4.0–7.0
Boron (B)	Yes	Suggested	$5–100 \times 10^{-3}$	0.1–1.0
Iron (Fe)	Yes	Yes	$50–150 \times 10^{-3}$	> 0.5
Manganese (Mn)	Yes	Yes	$10–20 \times 10^{-3}$	0.2–5.3
Copper (Cu)	Yes	Yes	$1–5 \times 10^{-3}$	$15–30 \times 10^{-3}$
Zinc (Zn)	Yes	Yes	$15–30 \times 10^{-3}$	$100–300 \times 10^{-3}$
Nickel (Ni)	Yes	Suggested	0.1×10^{-3}	$20–30 \times 10^{-3}$
Molybdenum (Mo)	Yes	Yes	$0.1–1.0 \times 10^{-3}$	1
Sodium (Na)	Beneficial	Yes	–	2–5
Selenium (Se)	Beneficial	Yes	–	$10–100 \times 10^{-3}$
Cobalt (Co)	Beneficial	Yes	–	$10–20 \times 10^{-3}$
Iodine (I)	–	Yes	–	$1–20 \times 10^{-3}$
Fluorine (F)	–	Suggested	–	0.1
Lithium (Li)	–	Suggested	–	$10–200 \times 10^{-3}$
Lead (Pb)	–	Suggested	–	$10–20 \times 10^{-3}$
Arsenic (As)	–	Suggested	–	$1–20 \times 10^{-3}$
Vanadium (V)	–	Suggested	–	$1–10 \times 10^{-3}$
Chromium (Cr)	–	Suggested	–	$1–2 \times 10^{-3}$
Silicon (Si)	Beneficial	Suggested	–	nd
Aluminium (Al)	Beneficial	–	–	$40–200 \times 10^{-3}$
Cadmium (Cd)	–	–	–	$5–10 \times 10^{-3}$
Mercury (Hg)	–	–	–	$2–5 \times 10^{-3}$

Essential elements for plants and animals are indicated. Mineral elements considered beneficial to plants, which improve the growth of various taxa under certain environmental conditions, are also indicated. Critical concentration for sufficiency is defined as concentration in a diagnostic tissue that allows a crop to achieve 90 per cent of its maximum yield. Critical concentration for toxicity is defined as concentration in a diagnostic tissue above which yield is decreased by more than 10 per cent. Data are taken from MacNicol and Beckett (1985), Brown et al. (1987), Marschner (1995), Mengel et al. (2001), White et al. (2004), and Pilon-Smits et al. (2009). It should be recognized that critical tissue concentrations depend upon the exact solute composition of the soil solution and can differ greatly both between and within plant species. The latter differences reflect both ancestral habitat and ecological strategies.
Source: White and Brown 2010.

major macronutrients for most plants are nitrogen (N), phosphorus (P) and potassium (K), with calcium (Ca), magnesium (Mg) and sulphur (S) also needed in smaller amounts. The two most limiting nutrients for agricultural yields are nitrogen and phosphorus, and these tend to be heavily applied. Many of the nutrients can be toxic in excessive amounts (Table 2.2).

2.4.4.1 Nitrogen

Nitrogen (N) is crucial for crop growth. It comprises nearly half of the dry weight of cell protoplasm and is involved in the formation of amino acids, cell division, production of carbohydrates, plant growth, and photosynthesis. Atmospheric nitrogen (N_2), which comprises approximately 78 per cent of air, is unavailable to plants. Plants use N in the form of ammonium (NH_4^+) or nitrate (NO_3^-). N in its available form is available naturally in the soil from weathered minerals, or is 'fixed' from the atmosphere by nitrogen-fixing bacteria which can form symbiotic relationships with plants. Microorganisms play a role in returning N to an inert state in a process known as denitrification. In agricultural systems, N is a limiting factor in crop production and is added as a fertilizer to maximize yields. Before synthetic fertilizers were available, N was added through the application of manures, and manure remains the main source of many subsistence low-yielding farming systems. In some rotational systems, legumes, which are known to fix N, are used to improve the soil; the use of legumes in this way is standard in organic systems. In the early industrial economies, concentrated manures such as guano were imported to meet N needs (Lawlor et al. 2001). The breakthrough of the Haber–Bosch process to artificially produce N was the driver for the increase in production in the twentieth century.

2.4.4.2 Phosphorus

Phosphorus (P) is also crucial for plant growth, particularly photosynthesis, respiration, energy transport, root formation and growth and seed formation, and it is a structural component of DNA. P is a scarce nutrient in natural and semi-natural habitats—the majority is locked in geological deposits—but it is also recycled from plant and animal waste and is added to soils as organic material such as farmyard manure, slurry, and green waste composts. Rock phosphorus can be applied, but this is a limited resource (Cordell et al. 2009).

2.4.4.3 Potassium

Potassium (K) is important for plant growth and determines fruit size, leaf thickness, and stem strength. It plays a major role as a catalyst in plant sap, in photosynthesis, transport of sugars, water and nutrient transport, transport of carbohydrates (plants that bear sugar-rich fruits have a high requirement for potassium), protein synthesis, and starch. Through regulation of water, K plays a role in protecting plants from drought (Zörb et al. 2014). The major sources of K are organic compost and manures and potash.

More information on nutrients and nutrient cycling in ecosystems can be found in Chapter 4 (Box 4.1).

2.5 Unproductive areas on farmland

Even in very intensive systems there are often areas of non-productive land interspersed with productive areas. As with the farming systems themselves, the scale and type of these alternative areas is highly variable, and this may depend on their purpose. Some of the most common non-cropped areas consist of semi-natural areas of woodland or scrub, boundary habitats such as hedgerows or other vegetative margins, waterbodies including ditches, rivers, streams, and ponds, and areas that have been planted on a temporary or semi-permanent basis but not for productive reasons. The latter may include cover (or catch) crops to reduce soil and water pollution, or specific plant mixtures designed to support wildlife such as farmland birds or pollinators. Some examples of non-productive areas are given in Box 2.2.

Box 2.2 Areas that are largely unproductive, i.e. that are not directly used to grow agricultural products, can still be managed to be 'useful'

Mitigation for pollution and erosion

Pollution and erosion can be mitigated on farmland with physical barriers (for example, buffer strips and cover crops). Riparian buffer strips are vegetative margins sown alongside rivers and waterbodies to prevent run-off of silt, nutrients, and pesticides from fields into watercourses. Cover crops are sown to protect the soil from wind erosion and improve soil fertility. In Europe they may comprise brassicas, legumes, or root crops such as forage turnip. In the African Sahel, a transitional area between the wooded Sudanian Savanna to the south and the Sahara to the north, 'fallow bands' are being established in cropped fields to prevent wind erosion. Fallow bands are shifting herbaceous windbreaks established by leaving 5-m-wide strips of land unmanaged during the rainy season, and maintained during the dry season to trap windblown soil particles. They are placed at right angles to the prevailing wind direction and moved windward across the field in each subsequent year. The process has been shown to control desertification and improve water management (Ikazaki et al. 2011).

Areas conserved for cultural purposes

Some areas on farms are left uncultivated to protect areas of high cultural value. In the UK, archaeological features such as burial mounds or relic medieval ridge and furrow systems are left unploughed. In India, sacred groves—patches of woodlands that are believed to have a spiritual significance—are conserved and often left as islands in the farmed landscape.

Continued

Box 2.2 *(Continued)*

Enhancement to support agricultural production

Non-productive areas can be used to enhance the agricultural system, even if production doesn't take place directly on them. A good example is the use of trees and shrubs in livestock systems, where they act as shelterbelts to protect livestock from adverse winter weather conditions in cool climates and provide them with shade in warmer climates. Some vegetative species also provide health benefits, which vary according to the chemical composition of the plants, and there is some evidence that livestock will self-medicate, depending on their health status. For example, an American study found that lambs with high parasite loads selected fodder with higher levels of tannins (a natural antiseptic) than their less-infected counterparts (Lisonbee et al. 2009).

Semi-natural areas/wildlife provision

Some areas on farmland are used specifically to preserve and enhance natural vegetation and the species that they support. This is common in Europe, where policy funding is provided to farmers to set-aside and manage land specifically for biodiversity. These areas may comprise field margins, such as conservation headlands, which are left unsprayed to encourage arable plants and insects; areas that are sown with non-crop plants to support pollinators and beneficial insects; or woodland to provide habitat for birds.

Alternative economic provision

There are frequently other economic enterprises carried out on farmland which may not be directly linked to food production but which shape the agricultural landscape. Recreational shooting on farmland can have a big impact through the management of woodlands and the introduction of 'game strips'. Game strips are areas of semi-natural habitats sown and managed to provide cover and insect food for game birds and mammals. In areas of high conservation value, non-productive areas may be enhanced for rural tourism by maintaining hedges and footpaths. Others may combine production with activities that encourage biodiversity, such as woodland management for timber production or charcoal burning. Increasingly, some farmers are seeking to diversify their farm income by using their land for multiple purposes and combining food production with opportunities for recreational activities such as shooting, or by providing light industrial units. However, willingness to diversify varies strongly. In Europe, young organic farmers were shown to be highly willing to diversify, while those managing intensive livestock holdings were unlikely to (Weltin et al. 2017). In Ghana the decision to diversify was linked to income, as only the wealthy had the facility to invest in diversification (Abdulai and CroleRees 2001).

Marginal areas

Some areas may be unproductive or difficult to access with machinery and these may be left unfarmed. These marginal areas may include stream margins, thin exposed soils, steep slopes, or other inaccessible areas. Some of these marginal areas may offer some of the other benefits outlined above.

2.6 Ecosystem impacts of agriculture

Management of land for production often conflicts with the natural biological processes that are taking place in agroecosystems. These potential ecosystem impacts of agricultural production have long been recognized. In the fourth century AD, Xenophon was calling for farmers to take care of soil, and both Plato (427–347 BC) and Aristotle (384–322 BC) argued that Bronze Age land use had degraded soil quality through deforestation (Montgomery 2007).

It is not just clearing of land for agriculture that causes problems; many of the inputs into the system do so too, particularly fertilizers and pesticides. The liberal use of artificial N has led to environmental pollution, N leaches easily from the soil and finds its way into watercourses where it is a pollutant and a contributor to the eutrophication of waterbodies (Steffen et al. 2015). In Europe between 11 and 31 per cent of N input is washed out of the system, much of it stored in ground water (de Vries et al. 2011). Similar issues occur with other nutrients. The solubility of P is influenced strongly by pH. Above a pH of 5, water-soluble components of P are found in the soil and may move across landscapes. In acidic conditions, P binds to soil particles but may still end up in watercourses due to agricultural soil erosion.

A wide range of unintended impacts have been associated with the use of chemical pesticides. Although some pesticides are biodegradable, others are more persistent and may not break down for months or years. This means that they require fewer applications, but over time they may move from the target area into the wider environment..

The negative impacts of herbicides on the agroecosystem are due to their fate after application, as they can enter watercourses and persist in the environment where they can impact non-target organisms, particularly invertebrates. Herbicides can also impact non-target organisms such as birds, by reducing seeds that are important food items. (Boatman et al. 2004). Increasingly there is concern about resistance of weeds to herbicides; for example, there is widespread resistance to glyphosphate in North America (Burke and Bell 2014). There are 220 plants known to be resistance to herbicides globally, although this is almost certainly under-reported (Heap 2014). This is a major challenge for agriculture in the future.

Insecticides suffer similar issues, with build-up of resistance and impact on non-target organisms. Neonicotinoid pesticides were once seen as relatively harmless. This was partly because they have lower toxicity to birds and mammals than to invertebrates, but there is now increasing evidence that non-target insects that feed on the nectar and pollen of plants (such as wild bees) can be negatively impacted in terms of behaviour and productivity (Cressey 2017; Woodcock et al. 2017).

Over-use of antibiotics in livestock systems can lead to resistance in the target animals, but is also thought to transmit resistance to other organisms, including humans. This can be via the food chain (meat in the retail chain has frequently been found to be contaminated with antibiotic-resistant bacteria) or by multiple routes of transmission, including by entering the environment though waste products, running off into waterbodies, and even in the air (Bonfoh et al. 2010).

Other biological issues connected with agriculture include soil degradation and biodiversity loss. Many of the issues are attributed to specialization and intensification, which are often seen as modern phenomena, but this is not the case. In his book *Dirt: The Erosion of Civilisations*, David Montgomery (Montgomery 2007) relates how in early Roman agriculture, small, hand-cultivated plots were used to support multistoried agriculture known as *cultura promiscua* which included grapes, olives, cereals, vegetables, and fodder. The high vegetation cover prevented soil erosion, deterred weeds, and created a microclimate which extended the growing season. The multiple root depths ensured little below-ground competition. With the advent of ox and plough, a single individual was able to prepare a much larger plot, but as a result, cropping was simplified and soil erosion increased. Around Rome the hillsides were deforested and shed sediment into the Tiber river which silted up to form the Pontine marshes which in turn became an infamous breeding ground for malaria and plagued the Roman population for centuries (Montgomery 2007).

Many of the biological conflicts between agricultural production and natural processes are explored further in later chapters of the book.

3
Biodiversity

3.1 Introduction

This chapter discusses biodiversity in different types of farming systems, including a consideration of the ecological importance of non-agricultural habitats within farm systems. The chapter outlines patterns of biodiversity, before exploring the relationship between agricultural production and biodiversity at local, regional, and global scales.

3.1.1 Biodiversity and agrobiodiversity

At its simplest, biodiversity, or biological diversity, is a measure of the genetic and phenotypic variability among living organisms. The term biodiversity can be used to describe diversity at a variety of scales: within species, between species, and even of ecosystems (www.cbd.int).

The Food and Agriculture Organization of the United Nations (FAO) defines agrobiodiversity as: 'The variety and variability of animals, plants and micro-organisms that are used directly or indirectly for food and agriculture, including crops, livestock, forestry and fisheries. It comprises the diversity of genetic resources (varieties, breeds) and species used for food, fodder, fibre, fuel and pharmaceuticals. It also includes the diversity of non-harvested species that support production (soil micro-organisms, predators, pollinators), and those in the wider environment that support agro-ecosystems (agricultural, pastoral, forest and aquatic) as well as the diversity of the agro-ecosystems' (FAO 2004).

3.2 Patterns of biodiversity within and between agroecosystems

In agroecosystems, there is a trend towards reduced biodiversity when compared with natural or semi-natural systems, particularly as agricultural management becomes more intensive. This is because increasing intensity in agricultural

The Biology of Agroecosystems. Nicola P. Randall and Barbara Smith, Oxford University Press (2020).
© Nicola P. Randall and Barbara Smith 2020. DOI: 10.1093/oso/9780198737520.001.0001

systems typically means lower complexity. Farmland is an artificial environment, and for the most part, modern agriculture is simplification by design; the land is managed to ensure one or few species (artificially selected and managed by humans) dominate, meaning that any others that occur do so in much lower numbers. Field enlargement, removal of semi-natural vegetation, and specialization all lead to systems dominated by monocultures (Frison et al. 2011), and it is largely this simplification that has led to a reduction in biodiversity (Landis 2017).

Despite these general trends, there are wide variations between farming systems, which fall along a gradient of complexity. At one of end of this gradient are the monocultures that dominate high-value cash crop areas, such as the wheat fields of Canada, the rice paddies of India, the cattle ranches of Texas in the USA, or the palm oil plantations of Indonesia. More diverse agroecosystems range from the mixed farms of crops and pasture in the UK and Switzerland, to areas of agroforestry such as the cork oak forests grazed by pigs in southern Spain, and rangelands or marginal lands that contain mainly natural vegetation but still support some livestock. The most diverse and complex include matrices of smallholder farms with mixed cropping, intercropped fields in developing nations, and, similarly, permaculture holdings in the developed world.

It is not only the number of species in a given area that is frequently altered in agroecosystems, but also their composition and distribution. Even if the total number of species stays the same, this is not usually considered the only factor important in measures of biodiversity. There are multiple methods to measure biodiversity (see Magurran 2004), but simple indices of diversity can be used to combine the total number of species (species richness) with evenness, to indicate the complexity of a community. This is illustrated in Figures 3.1a and 3.1b, which represent two systems, each with an identical number of individuals (100) and number of species (10) but with different levels of diversity, due to variations in the abundance of the different species represented. In Figure 3.1a, the species are spread fairly evenly, with no one species dominating. In comparison, the system illustrated in Figure 3.1b is much less diverse in comparison, with the population

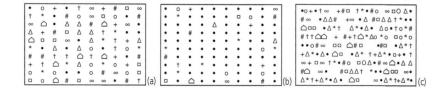

Figure 3.1 Biodiversity indices a, b, and c, where each square represents a given area and each symbol represents a different individual in that area. Each type of symbol represents a different species. Figures 3.1a and b have identical numbers of individuals (100) and species (10), but in (a) species are spread evenly, whereas in (b) one species dominates when compared with the others and so (b) would be considered more diverse using commonly used diversity indices. Figure 3.1c represents a more natural system, where spread of species and individuals is less uniform.

size of one species far exceeding that of the others. In managed agricultural systems, communities often shift from that illustrated in Figure 3.1a towards that shown in Figure 3.1b, although, in reality, in most natural systems the spread of individuals is also uneven, with patches and gaps in individuals and species as illustrated in Figure 3.1c. This again contrasts with agricultural systems, where croplands, in particular, have not only low diversity but also plants grown in densities that differ from naturally growing communities.

The management factors that influence biodiversity in agroecosystems vary between systems, but some key drivers are commonly found across multiple agroecosystems. Introducing new species into an agricultural system, or removing undesirable species, directly influences biodiversity on farmland, but there are various other management factors designed to support the species of agricultural interest that indirectly influence wider biodiversity (Baudron and Giller 2014). For example, the addition of nutrients, such as nitrogen and phosphorus, increases the productivity of target species, but may also reduce or eliminate other less competitive species, as they are unable to adapt to the new conditions or are simply crowded out. Some management actions can have wide-ranging effects, with implications far beyond the agroecosystems themselves: added nutrients and other chemicals may disperse from the farmed area and into the wider environment, impacting soil water and air quality elsewhere, and so indirectly affecting biodiversity; the removal of natural or semi-natural vegetation can directly impact the distribution and survival of species that depend on these habitats for part of their lifecycles; Emissions such as CO_2 and N_2O are thought to influence climate change, and so again indirectly influence wider biodiversity. Table 3.1 highlights some of the agricultural drivers of biodiversity change in and around agroecosystems.

Table 3.1 Some drivers of biodiversity change, and examples of their impacts in agroecosystems and more widely (adapted and developed from Baudron and Giller 2014)

Agricultural intervention		On-farm biodiversity impacts	Off-farm biodiversity impacts
Direct modifications of biodiversity			
Biotic additions	Domestic and/or exotic species	Negative impact on crops, native wild flora, and beneficial insects, through predation, disease, or competition. Reduction in farmland insects, birds, and mammals that rely on the affected native species for food	Extinction or reduced abundance of native species through predation, disease, or competition and a consequent negative impact on other species via trophic cascades
Biotic removals	Competitors, predators, pests, and parasites of domestic species	Increase in pest numbers due to removal of predators and parasites. Uneven farmland community due to dominance of a few species in absence of competition or predation	Impact on a wide range of species through trophic cascades due to removal of competition and predation

(Continued)

Table 3.1 *Continued*

Agricultural intervention		On-farm biodiversity impacts	Off-farm biodiversity impacts
Indirect modifications of biodiversity			
Biochemical cycles	Addition of nutrients (e.g. nitrogen and phosphorus)	Increases productivity of target species, but non-target species may not be able to compete. Favours a few species of weeds that then dominate and compete with the crop while reducing species that are important for biodiversity	Run-off into watercourses can reduce diversity of freshwater and marine habitats
Agricultural intensification	Increased mechanization; pesticides	Deep and frequent ploughing can destroy soil structure and reduce soil biodiversity. Pesticides can reduce non-target insects and lead to less resources for farmland birds and small mammals	Run-off due to poor soil structure and spray drift of pesticides can contaminate waterbodies and other habitats, thereby negatively affecting biodiversity
Land cover changes	Permanent (e.g. woodland removal) or temporary (e.g. annual cropping cycles resulting in bare ground at certain times during the year)	Habitat loss and structural simplification of landscape has negative consequences for farmland wildlife. Interruptions of natural lifecycles of organisms. Sowing times of annual crops, e.g. change from spring to autumn cereal sowing, reduces diversity of weeds and insects	Habitat fragmentation causes microclimate changes and other edge effects, which can impact on biodiversity of central areas; habitat may no longer be viable for species needing large ranges; communities may shift to species adapted to disturbance; connectivity in the landscape may reduce dispersal and genetic exchange between populations
Hydrological cycles	Irrigation and drainage; soil compaction	Irrigation and drainage can make fundamental habitat changes which shift the community composition. Soil compaction reduces soil biodiversity	Irrigation and drainage can restructure watercourses in surrounding habitat, leading to changes in biodiversity. Soil compaction can lead to run-off of soil into watercourses, leading to eutrophication and biodiversity loss
Carbon dioxide and nitrous oxide emissions	Climate change	Shifts in community composition of biodiversity	Shifts in community composition and range changes for both flora and fauna

The impact of agriculture can be considered even more widely, from within species genetics to global patterns, so it is useful to consider agrobiodiversity across all of these scales.

3.2.1 Global and temporal patterns of agrobiodiversity

Estimates of the number of species globally are debated, and range from 3 to 100 million, depending on the estimate used and how 'species' are categorized. For example, patterns within well-known taxonomic groups have put estimates of all eukaryotic species at 8.7 million (Mora et al. 2011). The number of species is constantly fluctuating due to speciation and extinction. Background extinctions are punctuated by phases when extinction rates increase dramatically. Probably the most well known of these mass extinctions is the Cretaceous–Tertiary mass extinction, during which time the dinosaurs (alongside many other organisms) died out. Many scientists believe that we have entered a new (human-induced) mass extinction (Ripple et al. 2017) and this is demonstrated in declines to biodiversity.

Global biodiversity declines are of increasing concern to conservationists and policy makers. The Millennium Ecosystem Assessment (2005) estimated that contemporary species extinctions could be up to 1,000 times higher than background extinction rates. Local species richness may have declined more than 10 per cent on average globally over the last 200 years (Newbold et al. 2015). Agriculture is thought to be one of the key drivers for biodiversity declines globally; intensification of practices within existing agricultural land is critical, but so is expansion into new areas and reductions in diversity of the goods produced (Lanz et al. 2018). The agricultural intensification that characterized the 'Green Revolution' of the 1960s saw the advent of intensive crop breeding, the increased use of fertilizers and pesticides, and a 40 per cent increase in the amount of land that is irrigated. The degradation of natural habitat via these routes and the quantity of land cleared has had a negative impact on ecosystem services such as pollination and natural pest control by beneficial insects, as their populations decline (Foley et al. 2005).

It is not only losses of habitat that drive changes in biodiversity, but also the fact that remaining natural biotopes have been fragmented. Fragmentation leads to reduced species richness, reduced movement of animals and plants between patches, and reduced nutrient retention (Wilson et al. 2016). It has been suggested that interactions between habitat fragmentation, climate change, and species invasions will inevitably lead to extinctions, as many species will be unable to move through fragmented habitats in order to reach areas with suitable climates and soils (Tilman et al. 2001).

Agricultural impacts on biodiversity are highly variable between regions, and are influenced by local environmental conditions, existing biodiversity, and the farming systems themselves.

It has been shown that agriculture is a risk factor for 80 per cent of all threatened terrestrial bird and mammal species (Tilman et al. 2017). The reduction of biodiversity in agricultural habitats since the mid-twentieth century is well documented in some regions, but monitoring and measuring biodiversity requires intensive effort, and for many groups of taxa, specialist knowledge is necessary to carry out identification. What is known about historic biodiversity losses on farmland is heavily skewed by what has been recorded, and a lack of long-term monitoring has hindered our understanding of population trends in many species and in many regions of the world.

Northern Europe, particularly the UK, is one of the regions of the world where monitoring of biodiversity, including in farmland, is most well established. In arable systems, for example, long-term declines in arable flora abundance and diversity have been demonstrated across Europe (Sutcliffe and Kay 2000; Potts et al. 2010), but this is not always consistent between species. A long-term monitoring study of farmland on the Sussex Downs in England demonstrated that there have been species losses and gains in arable weeds over a 38-year period, with rare species suffering the greatest losses (Potts et al. 2010). More recently, a meta-analysis of Europe-wide studies showed the same trend; although there have been some recent increases in generalists or species perceived as weeds, the average numbers of species overall declined, particularly the more rare arable flora (Richner et al. 2014). Due to their ability to compete with crops arable plants is often targeted directly via herbicide applications and indirectly through changes in management and use of fertilizers. The impact of this management, particularly sowing dates and herbicide use, has had variable effects on species depending on the lifecycle of the species in question. Wilson and King in their book *Arable Plants* (2003) asserted that rare arable plants now often only survive along the edges of fields where inputs are lower and the soil is less fertile, leading to less crop competition.

Farmland birds have also been studied at length in Europe, again particularly in the UK (see Box 3.1). Birds are considered a useful indicator of wider biodiversity trends as they need many different resources to support their lifecycles. These include: food, which may include a variety of seeds, berries, and invertebrates; shelter, meaning that a variety of structures are useful; and nesting habitats, which may be trees, shrubs, cropped areas, or grassland, depending on the species. Many birds need a combination of different habitats, and their requirements vary depending on their lifestage and the time of year. Figure 3.2 illustrates the resources needed to support birds on lowland farmland in northern Europe.

Although no other animal groups are as well recorded as birds, butterfly numbers have been monitored in the UK since 1976 and also show consistent declines. Reductions in UK butterfly abundance were reported to be as much as 76 per cent for habitat specialists and 57 per cent for species of the wider countryside in 2016 when compared with the 1976 figures (Brereton et al. 2017). The European Environment Agency has attributed this decline to the intensification of agriculture and an associated failure to manage grasslands; other studies have identified pesticides as an important factor. Concerns about pollinating insects, particularly bees, have led to analysis of existing datasets. These indicate that agriculture is having a negative effect on pollinator populations both globally and locally on farmland via habitat loss and pesticide use (Potts et al. 2010).

Spring/summer

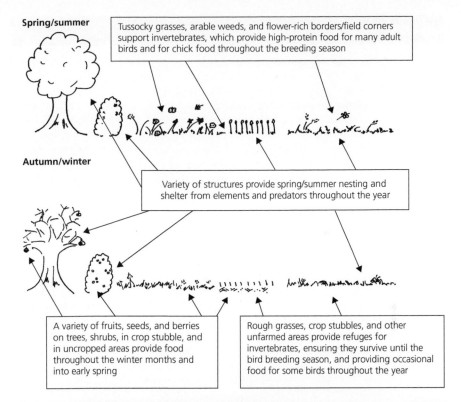

Tussocky grasses, arable weeds, and flower-rich borders/field corners support invertebrates, which provide high-protein food for many adult birds and for chick food throughout the breeding season

Autumn/winter

Variety of structures provide spring/summer nesting and shelter from elements and predators throughout the year

A variety of fruits, seeds, and berries on trees, shrubs, in crop stubble, and in uncropped areas provide food throughout the winter months and into early spring

Rough grasses, crop stubbles, and other unfarmed areas provide refuges for invertebrates, ensuring they survive until the bird breeding season, and providing occasional food for some birds throughout the year

Figure 3.2 Resources used to support birds on lowland farmland in northern Europe. The multiple resources needed by farmland birds throughout the year explains why they make good indicators of wider on-farm biodiversity. Original line drawings: Nicola Randall.

Plate 3 figures a and b illustrate current global biodiversity and agricultural intensity respectively. It can be seen that high agricultural intensity often correlates with low biodiversity. For example, parts of the northern USA currently have high levels of cropping density and corresponding low diversity, but these direct comparisons can be misleading. In regions where agriculture is fairly stable, the relative rate of change in biodiversity is often lower than in regions where contemporary agriculture is rapidly expanding or intensifying. The USA has a high rate of historic extinctions, and although the level of biodiversity is relatively low in comparison to many other regions, the current rate of change is likely to be slower, as the USA already has many established areas of large-scale intensive farming, and the state of biodiversity reflects these historic losses.

Regions with high biodiversity and high-intensity farming (Plate 3) are the ones that may be at greatest risk of high biodiversity losses in the near future, particularly where there is a high rate of agricultural change. Many of the most diverse regions are tropical and the native vegetation is often forest. Biodiversity losses in tropical regions are often largely due to the expansion of agriculture into these forests. Between 1980 and 2000, over 80 per cent of the land converted for agricultural expansion was previously forested (Gibbs et al. 2010).

Box 3.1 Bird monitoring in the UK

One of the oldest bird monitoring generalists was established in the UK. Since 1994 the British Trust for Ornithology (BTO) has coordinated volunteer surveys of breeding birds, and analysis of these results, coupled with earlier, less systematic surveys, has allowed researchers to investigate the impact of agricultural practice on their populations. To estimate the state of farmland bird populations, the UK Government has recognized 19 farmland bird species categorized as those '*species feeding in open farmland during the breeding season, even though they may nest in woods and hedges*'. It is known as the Farmland Bird Index. In Europe similar indicators have now been developed as part of the Pan-European Common Bird Monitoring Scheme, which allows comparison between countries.

Between 1970 and 2015 the UK Farmland Bird Index halved, with most of the declines occurring during the late 1970s and the 1980s when there were rapid changes in farmland management. Although the rate of decline has slowed, there was still a further decline of 8 per cent between 2009 and 2014, which is mostly accounted for by farmland bird specialists that are more vulnerable to changes in agricultural management, and this decline appears to be continuing (Figure 3.3). A European analysis demonstrated that agricultural intensification, centred around a push for increased yields, particularly in cereals, has driven a similar decline across Europe (Donald et al. 2001). These changes in farmland bird populations are reflected in other taxa for which there are records.

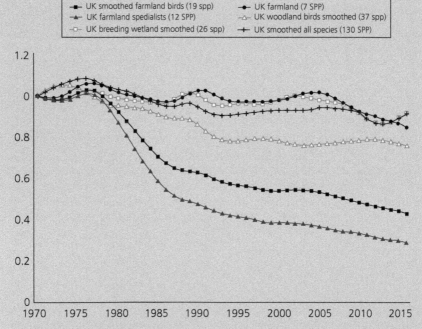

Figure 3.3 Relative change in average population size of terrestrial birds in different habitats in the UK since 1970 using smoothed data. The smooth farmland bird data are an amalgamation of farmland bird generalists and specialists.

Source: Data from BTO. Used with permission.

The Philippines is one of the most biodiverse countries in the world, but is also one of the regions suffering from high biodiversity losses, with more than 700 threatened species. Land pressures in the Philippines are varied and include mining and deforestation, but agriculture is also a big driver. Nearly 42 per cent of the land area in the Philippines is classed as agricultural (based on 2016 FAO estimates: http://www.fao.org/countryprofiles/index/en/?iso3=PHL). Traditionally, Filipino agriculture has been quite diverse, combining cultivated areas with permanent crops, but also involving swidden agriculture (agriculture that involves cutting and burning or plant growth followed by cultivation and cropping for a few years, followed by a period of fallow to allow degraded soil to regenerate). A combination of factors surrounding agricultural management have led to biodiversity losses: the secondary forest produced by this Swidden agriculture is less diverse than the original forest; many of the species used for farming are exotic, so support less native biota; fallow periods are becoming shorter as food production requirements increase; and the diversity of agricultural products is decreasing as exports become more important to the economy. As agriculture in the region intensifies, biodiversity is likely to come under increasing pressure.

The Philippines contrasts with Malaysia, which again has high diversity but has a much lower cropping density. In actual fact, less than 25 per cent of the total land area in Malaysia is currently classed as agricultural (compared with nearly 45 per cent in the USA and 42 per cent in the Philippines), the agriculture overall is less intensive, and Malaysia has one of the highest levels of environmental regulation in South East Asia, which may help protect biodiversity into the future. Despite this, much of Malaysian native biodiversity is still at risk. Around 60 per cent of the current agricultural area is devoted to oil palm production, which is known to have much lower biodiversity levels than unconverted forest due, in part, to the uniformity in tree age structure and species of plantations (Fitzherbert et al. 2008).

Malaysia had the highest recorded density of threatened species due to palm oil production in 2005 (Turner et al. 2008), and expanding plantations continue to replace natural forests, leading to losses of biodiversity, including large mammals and forest birds (Shevade and Lobada 2019).

Dryland areas have historically been considered less threatened by biodiversity losses than tropical regions. Dryland regions tend to have comparatively lower underlying biodiversity, and in some ways appear to be more resilient to change, yet these areas are also at great risk of biodiversity declines. Southern Africa, for example, has high levels of floristic diversity, but increases in domestic grazing cause changes to the vegetation and can lead to losses of endemic species and of biodiversity more widely in arid and semi-arid grasslands. In other parts of Southern Africa, technological developments have enabled more marginal areas to be utilized for agriculture, including arable farming in dryland areas, which are unlikely to ever be able to revert to their original vegetation (Darkoh 2003).

It is predicted that tropical and dryland regions in South America and Sub-Saharan Africa are likely to incur the highest rate of agricultural conversion into the future (Laurance et al. 2014), and therefore these are the regions where future agriculture is likely to impact on biodiversity.

3.2.2 Landscape-scale agrobiodiversity

The landscape in which farms are situated impacts on the biodiversity of the agricultural area in question, driven by the underlying topography, soil type, water availability, climate, and agricultural practice, as well as the surrounding land use. This is illustrated in Figure 3.4. Figure 3.4a shows prime flat farmland in Oregon, USA, which supports an intensive cropping system, whereas Figure 3.4b shows farmland in Wales, UK, where the diverse topography and thinner soils on hillsides limit farming options to lower intensity livestock and forest products.

Figure 3.4 Contrasting landscape diversity, where Figure 3.5a shows intensive arable farming on prime flat farmland in Oregon, USA, and Figure 3.5b shows low-intensity mixed livestock farming and forestry on thin upland soils in Wales, UK. Photos: Barbara Smith

Variations in habitat fragmentation, semi-natural areas, the amount of farmed landscape, farming practices, and intensity on the different farms in the region all impact on the biodiversity within each individual farm as well as on the landscape as a whole. A challenge associated with the intensification of farming is the resulting fragmentation of wild habitats where the non-farmed remnants can be not only small but also unconnected, so that species may be unable to (1) maintain viable populations due to lack of food and refuge and (2) move to a different patch due to a lack of connectivity between appropriate habitats. If there is little immigration and emigration between these small, isolated metapopulations, they can sometimes fall below their minimum viable population size, with populations that are too small to bounce back from events such as disease outbreaks or extreme weather events, and have limited gene exchange and no chance of range expansion. Habitat fragmentation, with little connectivity of suitable habitat, leads to vulnerable isolated populations and contributes to biodiversity decline. Batary et al. 2010 found that pollinators could be enhanced more effectively in simple agricultural landscapes than in more complex landscapes by introducing agrienvironmental management, probably because the more complex landscapes already contain refuges and corridors. There are significant efforts to introduce wildlife corridors across agricultural land, but experts debate the best way to do this, and the landscape itself may influence the success of conservation measures. Box 3.2 outlines some of the ecological theories that can be used to inform biodiversity conservation on agricultural land.

Box 3.2 Ecological theories that have been used to inform conservation on and around agroecosystems

In 1967, MacArthur and Wilson's theory of island biogeography identified that islands surrounded by sea are similar to islands surrounded by hostile habitat such as a semi-natural biotope on farmland surrounded by intensive agriculture. The number of species on an island will reach an equilibrium based on the rate of extinction on the island and the rate of immigration to and from islands. Larger islands support more individuals or species, as do islands that are closer to a mainland (another area of similar semi-natural habitat), and linked islands (or biotopes) will support more species than unlinked islands, but arguments still rage as to how this should best be achieved. In a seminal paper, Rhys Green and colleagues (Green et al. 2005) argued that the management of land for both agricultural production and conservation is essentially a choice between land sharing and land sparing. Land sparing represents a scenario where land use is divided so that it is either managed intensively for wildlife or, separately, managed intensively for agriculture. In contrast, land sharing farmland is an amalgam of productive agricultural land and semi-natural areas, where wildlife co-exists with the extensive agriculture.

The debate on which approach is more sustainable and leads to resilient productive systems is ongoing (Fischer et al. 2014). The two approaches will support very different types of biodiversity. Land sparing options, which allow for nature reserves,

Continued

Box 3.2 (*Continued*)

have greater potential to support the characteristic species associated with undisturbed habitats; however, viable populations are only maintained in areas of sufficient size, with small scraps of land unlikely to support resilient communities. Examples of land sparing can be seen in the USA, where national parks are preserved as wilderness while there is intensive production elsewhere on fields as large as 250 ha, supporting minimal biodiversity. In contrast, traditional farming systems of Indonesia follow a more land sharing approach. Agriculture is interspersed with more natural areas that frequently provide ecosystem services via insect pollinators and pest enemies, or they may act as buffer zones beside watercourses. However, apart from a few farmland specialists, the majority of the species supported are often generalists and their populations are vulnerable to extinction should the agriculture become more intensive.

Much of the ongoing debate comparing the merits of land sharing and land sparing to deliver benefits for both production and biodiversity considers the scale of implementation of each. Land sparing could be implemented at regional, farm, or field levels, as might land sharing, where farmers practise intercropping or under-sowing that supports biodiversity and where arable plants or 'weeds' may be tolerated, and this scale may influence the ability of each to deliver. This issue of scale is illustrated in Figure 3.5.

In reality, how land is either shared or spared will be context dependent, depending on land availability and population density as well as agricultural policy.

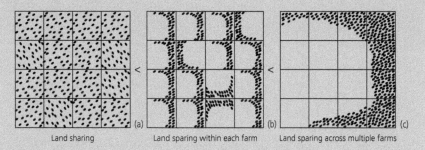

Land sharing   Land sparing within each farm   Land sparing across multiple farms

Figure 3.5 Schematic summarizing the landsparing, landsharing continuum where within-farm 'biodiversity-friendly' farming (a) is compared with landscapes that involve land sparing within large farms (b) or across a group of farms (c). In each landscape, the same total area (denoted by black dots) is given over to wild nature.

Source: Balmford et al. 2012.

3.2.3 Farm-scale agrobiodiversity

Agrobiodiversity within farms can be as variable as agrobiodiversity between farms. In addition to the productive areas within agricultural land, there are often many areas that fall within the agroecosystem but are not used for production (see Box 2.2, Chapter 2). These areas may be relatively small (e.g. consisting of field margins and boundaries), they may be unused because they are inaccessible, or they may be purposefully left as unproductive areas to provide other uses such as shade for livestock or for cultural reasons.

The size and nature of these non-productive areas can be highly variable and their presence, design, and location can all influence farmland biodiversity. In studies of temperate farm woodlands, Fitzgibbon (1997) found that even small farm woodlands are important refuges for bank voles and wood mice, especially following harvest. He also found that woodlands connected by hedgerows supported higher densities of these small mammals, but that population increases were greater from spring to autumn in isolated woodlands (possibly because populations could not disperse easily). Woodland structure and surrounding crops were both important drivers of populations within the woodlands. In a similar type of study, Usher et al. (1993) investigated invertebrates in new woodland and found that there were likely trade-offs between maximizing core woodland area (more characteristic woodland species) and having longer, thinner woodlands (higher overall species richness). But they found that taxonomic groups behave differently. For example, ground beetles were most impacted by woodland shape and spiders by woodland isolation.

Different types of crops on a farm also support more or less biodiversity. Crop type influences biodiversity through: (1) the direct provision of resources such as pollen and nectar for flower and seed feeders, such as invertebrates and birds; (2) its structure, by providing access, refuge, and nesting resources (Morris et al. 2004); and (3) and its associated management—for example, leaving cereal stubble over winter has been shown to increase the productivity of breeding farmland birds (Gillings et al. 2005), and pesticide use reduces insect diversity (McLaughlin and Mineau 1995).

The extent to which pasture supports biodiversity depends largely on the intensity of its management and the composition of the landscape in which it sits. Floristically rich grasslands support significantly more insect diversity than intensive pasture (Siemann et al. 1998) and the regulation of grazing has been shown to increase biodiversity under lighter grazing regimes (Plantureux et al. 2005).

3.2.4 Species and genetic agrobiodiversity

It is not only the underlying biodiversity of the non-farmed species that has been impacted by agricultural developments, but also the agricultural species themselves. The world has more than 50,000 edible plants, yet 90 per cent of the world's energy demands are met by just 15 crops and two-thirds of our calorie intake is provided by only three species: rice, maize, and wheat. Although the species used within agroecosystems is thus limited, new varieties of these species are continuously being bred for production. The FAO recognized over 8,000 different breeds of livestock (those with similar phenotypes and characteristics) in 2010, each bred to deal with specific productivity requirements, diseases, and climatic and other local conditions (Hoffman 2011).

Despite this, the process of domestication of crops and livestock has led to a reduction in their genetic diversity, and most agricultural systems utilize few of the varieties available. Early agricultural practice involved the selection of desirable traits in wild species which were eventually domesticated, and these early processes have left a genetic signature in the genome of modern crops. As early farmers selected plants from relatively few individuals from the progenitor species, much of the original diversity was lost. Furthermore, only the 'best' plants formed the next generation. This process led to a genetic bottleneck where genes that influence the desirable traits underwent a severe loss of diversity due to selection. Neutral genes, those that are independent of those traits, also lost diversity but to a lesser extent. In the case of neutral genes the scale of loss is related to the size of the population and the duration of the bottleneck (Doebley et al. 2006).

A negative consequence of selective breeding is inbreeding depression for traits showing dominance, which may lead to a loss of genetic diversity as well as having negative impacts on health. Contemporary commercial agricultural practices are aimed at producing high-yielding crops that have uniform characteristics for ease of on-farm and post-production management, such as machine harvesting in large fields (Esquinas-Alcázar 2005). Over time, plant breeding for increased yield has led to a loss of beneficial traits such as pest resistance as there are often trade-offs between traits that encourage production and other traits. This, in turn, has led to dependence on increasingly strong pesticides (Gruber 2017). Denison (2012) argues that although many of these trade-offs will have probably occurred at some point due to natural selection, without selective breeding, traits that encourage production at the expense of other desirable factors would have been rejected as part of natural evolution. This reliance on a few high-yielding varieties has led not just to a narrow genetic base for the crops themselves but to a simplification of farming systems where a few crops tend to be grown in large fields, managed for pests and disease through high chemical inputs.

An analysis of global data from the FAO has shown that food supplies across the world have become more similar in composition. While national food supplies have, on average, become more diverse, there is more similarity between countries. Globally, the same crops are driving this similarity between nations, i.e. wheat, rice, soybeans, palm oil, and sunflowers (Khoury et al. 2014). At the national level, the FAO, in a stocktake of global crop genetic diversity, showed that local genetic diversity is being eroded due to local varieties being lost in modern production systems (as reported in Lanz et al. 2018). In livestock, a great deal of research is undertaken to understand how to manage populations of animals to minimize inbreeding (Kristensen and Sørensen 2005).

In 2015, the European Commission introduced a 'three crop rule' where in countries that are European member states, farms over 30 ha are required to grow at least three crops, and this was introduced under a banner of diversification. That three crops is considered diverse underlines the loss of both genetic and species diversity in agricultural fields, particularly in the areas growing those dominant crops.

3.3 The importance of biodiversity in agricultural systems

3.3.1 Ecosystem functioning and provision of ecosystem services

Biodiversity is important in ecosystems, because each species contributes to whole ecosystem functioning in a slightly different way, and these varying contributions impact on ecosystem properties and processes (Wood et al. 2015). The role species play in a system varies according to their function and their relationship with other species. Species can be said to have 'functional traits', which are defined as 'those that influence ecosystem properties or species responses to environmental conditions' (Hooper 2005). A more diverse system is likely to have a higher diversity of traits, which increases ecosystem functioning (this is explored further in Chapter 4). This generally means that complex systems are also stable, as they are more resilient to change than simpler systems.

3.3.2 Stability and resilience

Holling (1973) defined resilience as a measure of the persistence of systems and of their ability to absorb change and disturbance and still maintain the same relationships between populations or state variables. Walker et al. (2004) developed this term to describe 'the capacity of a system to absorb disturbance and reorganize while undergoing change so as to retain essentially the same function, structure, identity, and feedbacks'; in other words, the system can absorb shocks and bounce back rapidly. Some authors separate resilience and stability, with a stable ecosystem defined as one that deviates little from its natural state despite environmental variations, and resilience defined as the ability of an ecosystem to return to its original state following disturbance (Cleland 2011). Biodiversity can impact on both ecosystem stability and resilience.

Stability and resilience are important for agriculture as they both impact productivity. Low year-on-year variance in crop yield is important for food production and is an important measure of stability in agricultural systems. Variability can be high in agroecosystems, but experimental studies and large-scale field trials have shown that agricultural biodiversity can reduce variance (Frison et al. 2011). Resilience to extreme weather events such as drought or extremes of temperature is also important for both ecosystem stability and agricultural productivity. Again, biodiversity can help support this resilience; having multiple species makes it more likely that at least some of them are less vulnerable to specific stressors, thereby resulting in functional resilience, i.e. if one species fails due to poor weather conditions, another species could provide the services potentially lost in that year.

In a study by Thorlakson and Neufeldt (2012), subsistence farmers in Kenya were provided with a variety of tree species in order to help them diversify their farming systems. Farmers that took part in the programme were found to be more self-sufficient than farmers that didn't take part during drought and flood events due to the increased diversity of the systems and of their products.

Sometimes the loss of even a single species may have seemingly disproportionate impacts on the system if no other species perform the same function. These 'keystone' species may impact on multiple species or on an essential part of the system, and manipulation of their populations may have unexpected effects. An example of this is shown in Box 3.3. In other agroecosystems the functional value of specific species may be less well defined but still important for the system as a whole. Arable plants, for example, provide food and shelter for higher taxa. It is known that weed seeds are an important food resource for some species of insects, birds, and small mammals (Tooley and Brust 2002; Westerman et al. 2003; Holmes and Froud-Williams 2005; Holland et al. 2006). Seeds are also consumed by invertebrates, commonly by some carabid species (Tooley and Brust 2002; Petit et al. 2014) and ants (Pearson et al. 2014), for example, and these two invertebrate taxa are important dietary items for farmland bird chicks and adults (Holland et al. 2006).

Box 3.3 Sparrows in China

In 1958, Mao Zedong introduced his 'Four Pests' campaign into China. The aim was to remove pests that transmitted disease and those that compromised food production. Mao selected mosquitos, rodents, other flies, and, significantly, sparrows, which foraged in crop fields and ate grain. Mao encouraged his citizens to destroy sparrow nests, to bang pots and pans to prevent the birds from settling on trees and resting so that they eventually died from exhaustion, or to simply shoot them. A contemporary issue of the US magazine *Time* reported that: 'At dawn one day last week, the slaughter of the sparrows in Peking began, continuing a campaign that has been going on in the countryside for months. The objection to the sparrows is that, like the rest of China's inhabitants, they are hungry. They are accused of pecking away at supplies in warehouses and in paddy fields at an officially estimated rate of four pounds of grain per sparrow per year. And so divisions of soldiers have been deployed through Peking streets, their footfalls muffled by rubber-soled sneakers. Students and civil servants in high-collared tunics, and schoolchildren carrying pots and pans, ladles and spoons, quietly took up their stations. The total force, according to Radio Peking, numbered 3,000,000.' (*Time*, May 5, 1958).

The campaign was a success and led to the near extinction of the Eurasian tree sparrow. However, the unforeseen consequences were devastating. Sparrows didn't just eat grain; they also acted as a pest control agent by feeding insects to their young. Rice yields were recorded as decreasing after the campaign, while pest populations increased to devastating levels in an agricultural landscape that was already under stress from deforestation and misuse of pesticides (Shapiro 2001), contributing to the complex set of conditions that led to the Great Chinese Famine of 1959–61. Too late, in 1960, the campaign against the sparrows was halted, leaving an early example of how removing a key species can have unanticipated and negative consequences.

A reduced genetic pool means that all individuals are similar, and thus more vulnerable to stochastic events such as outbreaks of pathogens or periodic events such as drought or pest outbreaks. There are many well-documented examples of disease outbreaks that have been caused by loss of resistance in crops with narrow genetic bases. This was illustrated in 1970 in the southern USA, where a lack of resistance due to inbreeding caused over half of the standing maize crop to be lost to a fungus outbreak of *Helminthosporium maydis* (Esquinas-Alcázar 2005). Genetic diversity can provide resilience in the face of these events, and many local or wild varieties once considered of no use to commercial production have since been found to have important genetic traits to fight against pests, diseases, and adverse environments. These species are now being used as donors in improving crop breeding to deal with insect and disease resistance (Esquinas-Alcázar 2005).

3.3.3 Productivity

Biodiversity can also have positive effects on productivity, particularly in temperate grasslands. The relationship between biodiversity and primary production was shown in eleven long-term grassland experiments in the northeastern USA, where grasslands containing multiple species had increased biomass production when compared with monocultures (Tilman et al. 2012). Similarly, in the UK, when diverse grasslands were compared with simple grasslands they showed a 43 per cent increase in hay production (Frison et al. 2011). These biodiversity effects in the wild are common and strong key drivers of productivity, but the effects have been shown to be stronger in nature than in experiments (Duffy et al. 2017), and there is some indication that these effects are less clear in some agricultural systems. This is discussed further in Chapter 4.

3.4 Issues and challenges for management of biodiversity in and around agroecosystems

Agricultural systems are unnatural systems. They are managed for a specific purpose, the production of one or more specific commodities. Unfortunately this management usually directly conflicts with the enhancement of biodiversity. Even in very mixed farming systems, some species are being encouraged and managed, which may be to the detriment of others that are of no economic interest. In many systems, particularly contemporary commercial systems, this pattern is extreme: the focus is on the enhancement of very few species, whether crops or livestock. Usually, native vegetation and corresponding animals have to be removed to create the space for the species of agricultural concern, such as in planting a cereal field.

This simplification is the first and most direct trade-off between agricultural production and biodiversity, but, as discussed throughout this chapter, many other aspects of management of agroecosystems create direct conflict between biodiversity and agricultural production. The application of fertilizers in crop or grassland systems, for example, increases the productivity of the target plant species, but this is to the detriment of other species that are unable to compete in a high-nutrient environment; in a study of over 1,500 grassland sites in Germany, Socher et al. (2013) found that on average plant species richness was reduced by around 15 per cent in fertilized grasslands. Other management tools such as the application of herbicides are designed specifically to reduce competition from non-desirable species, but there are also aspects of agriculture that impact biodiversity but are due more to other factors of the management than direct trade-offs with production. Examples of this include when nutrients or pesticides inadvertently escape the agroecosystem into natural or semi-natural areas. See Table 3.1.

3.5 Tools for sustainable management of biodiversity in agroecosystems

The *conventional intensification* of agriculture, through high agrochemical inputs (insecticides, herbicides, and fertilizers) and monocultures, has enhanced production over the last few decades, but is beginning to fail to track demand as yield increases slow. Because agricultural land is limited, many experts are calling for additional production to come from *sustainable intensification* of agriculture, which aims to produce more from the same or a smaller area of cultivated land, and with reduced inputs while minimizing adverse environmental impacts. Sustainable intensification includes several complementary approaches to intensification such as: integrating genetics with husbandry (e.g. novel breeding tools); precision farming; and managing ecosystem services underpinning production (e.g. natural pest regulation and pollination). The latter is an emerging pathway, known as *ecological intensification*, which is a knowledge-intensive process that requires optimal management of nature's ecological functions and biodiversity to improve agricultural system performance, efficiency, and farmers' livelihoods. Chapter 8 outlines a variety of options for more sustainable agroecosystems.

4

Biotic Interactions and Ecosystem Functioning

4.1 Introduction

This chapter introduces different functional groups and their role in agroecosystems and in the provision of ecosystem services. As discussed in Chapter 3, biodiversity can offer a variety of benefits or 'services' to natural systems including agroecosystems, but it is, in fact, the biotic interactions that take place within and between the species in the systems that provide these 'ecosystem services'. The biological functions of living organisms all contribute to natural ecosystem processes in some way and these relationships are discussed in the following sections.

4.2 Ecosystem services

Ecosystem services are benefits that ecosystems provide (to humans) (Costanza et al. 1997), although this may be indirectly. They can be categorized into four broad types:

- Provisioning—the products that come from ecosystems, such as food and water
- Supporting—those that are necessary for the production of other services—such as soil formation and nutrient cycling
- Regulating—the regulation of the processes and the benefits that come from them, such as air and water purification, or pest and disease control
- Cultural—the non-material benefits that people obtain from ecosystems; these may include a sense of wellbeing, and recreational and educational experiences

(Millennium Ecosystem Assessment 2005).

4.2.1 Ecosystem services in agricultural ecosystems

Agricultural systems are primarily aimed at the enhancement of provisioning services—namely yield of a crop or support of livestock—but this can be at the cost

The Biology of Agroecosystems. Nicola P. Randall and Barbara Smith, Oxford University Press (2020).
© Nicola P. Randall and Barbara Smith 2020. DOI: 10.1093/oso/9780198737520.001.0001

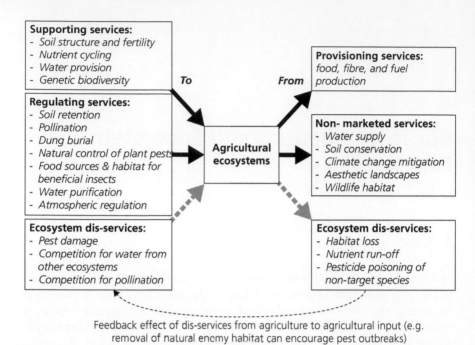

Feedback effect of dis-services from agriculture to agricultural input (e.g. removal of natural enemy habitat can encourage pest outbreaks)

Figure 4.1 Key examples Reprinted from: Zhang, W., Ricketts, T.H., Kremen, C., Carney, K. and Swinton, S.M., 2007. Ecosystem services and dis-services to agriculture. Ecological economics, 64(2), pp.253–260, with permission from Elsevier. of ecosystem services and dis-services that are particularly relevant to agricultural ecosystems.

of other services, and may even negatively impact on wider ecosystems or, in other words, provide ecosystem dis-services (Figure 4.1; Zhang et al. 2007b). Yet research shows that all ecosystem services, particularly supporting and regulating services, are essential in order to effectively deliver the provisioning services needed in crop productive systems (Bommarco et al. 2013). This potential trade-off in agroecosystems is often managed through external inputs into the systems, e.g. applications of artificial pesticides and additions of chemical fertilizers.

There are, however, natural interactions that take place within ecosystems that support the delivery of these ecosystem services. Many of these are delivered by specific biological features or 'traits' of biological organisms.

4.3 Functional traits

Traits are usually considered to be defining characteristics of an individual or species. In biological terms, functional traits are those that 'influence ecosystem properties or species responses to environmental conditions' (Hooper et al. 2005). An understanding of these functional traits and their influence on the wider processes is important to more effectively understand agroecosystems. We can also use this understanding of natural biotic functions in order to manage the different types of services, and make decisions about any trade-offs that may have to be made between them (e.g. between yield and biodiversity see Chapter 3).

4.4 The value of different functional groups for delivering ecosystem services in agricultural systems

It is generally understood that more diverse ecosystems support higher levels of ecosystem function, although in the mid-1990s scientists began to understand that it is actually the characteristics of the different species in the system that provide the functions rather than the species themselves (Grime 1997). Species abundance and diversity vary within and between different biotopes, and a selection of complex interactions continually takes place between different organisms (biotic interactions) and between organisms and the wider (abiotic) environment. Increasing the diversity of traits increases ecosystem functioning, as there is scope for greater complementarity in organisms' use of resources (resource partitioning) and in how they modify their surrounding environment in ways that impact other species (known as facilitation). In simple terms, species differ in the ways in which they contribute to ecosystems (Wood et al. 2015).

Stable natural or semi-natural communities tend not to be dominated by one or two species. Management of land for food production often means that systems become (to a greater or lesser extent) simplified, as target plants or animals are encouraged to thrive at the expense of others. This is particularly the case where one output dominates the system, such as in many intensive monocropped systems or high-production livestock systems, but even in traditionally more diverse systems, such as smallholder mixed cropping or home garden systems, there is an increasing trend towards simplification and intensification. This simplification of natural systems, often combined with other management factors, influences the nature of the biotic (and abiotic) interactions that take place within and around agricultural systems, and may result in instability (see Chapter 3, section 3.3.2), although it is often difficult to predict exactly what the impacts will be. Shifts in species compositions and even reduced species richness do not necessarily mean that important ecosystem functions are also lost. Some processes are provided by multiple organisms, and where this is the case there is an element of 'functional redundancy'. In contrast, where a few specialized taxa provide specific processes, even a small change to the species composition could impact on the wider ecosystem (Bommarco et al. 2013).

According to Wood et al. (2015), 'It is not necessarily the diversity of species, but rather the diversity of functional traits that is important for ecosystem functioning.'

As the global trend for intensification of agriculture continues, an understanding of biotic functions, the way in which they support essential ecosystems services within and around agricultural environments, and which functions are most vulnerable to change is becoming increasingly important for underpinning future decision making.

4.5 Key ecosystem services in agroecosystems and the organisms that deliver them

The next sections describe, in more detail, some of the key biotic interactions within agroecosystems, together with the ecosystem services they deliver and potential methods for their enhancement.

4.5.1 Carbon and nutrient cycling

Nutrients and nutrient cycling are essential in ecosystems. Carbon, hydrogen, and oxygen are essential to life and part of living cells, but a variety of other nutrients are essential to the functioning of life. Essential nutrients are those that organisms need in order to complete their lifecycles. In natural systems, nutrients in the environment are utilized and recycled around both the biotic and abiotic components of the systems. Nutrients come in two forms: 'organic' and 'inorganic'. Organic compounds contain carbon and are contained in living or remnants of living organisms. Carbohydrates, lipids, and proteins are examples of organic nutrients.

Plants are autotrophs and can produce complex organic compounds themselves, deriving cell carbon from atmospheric carbon dioxide (CO_2). Most plants do this via photosynthesis (a process during which light energy is captured and used to convert water and carbon dioxide from the air into oxygen and the organic compounds (see Box 4.1 for more details), but plants also require inorganic nutrients (such as nitrogen, phosphorus, and potassium) to enable them to do this. Plants cannot access nutrients in organic forms. One of the main ways in which plants obtain essential inorganic nutrients is in an aqueous ionic form via their roots. Table 2.2 in Chapter 2 lists the essential inorganic plant nutrients.

Animals and other heterotrophs need to feed on other organic organisms in order to obtain their nutrients, as they cannot manufacture them from inorganic nutrients. This natural relationship between autotrophs and heterotrophs can be simplified within the carbon cycle, whereby autotrophs such as plants manufacture organic compounds, these heterotrophs consume the plants, the organic compounds are broken down during the process of respiration, and this releases CO_2 back to the atmosphere to be used again in photosynthesis.

In agricultural systems, nutrients are removed from the system as agricultural products are taken away, but even where the remains of living organisms form organic matter, it is essential that this organic matter in the soil is broken down or

Box 4.1 Photosynthesis in plants: C3, C4, and CAM pathways

Autotrophs are organisms that can produce complex organic molecules from inorganic materials. Some bacteria do this through a process called chemosynthesis, whereby inorganic materials are oxidized during a process of specialist respiration. Other bacteria and all green plants do this through a different process that requires light energy, called photosynthesis.

Plants convert atmospheric CO_2 (carbon fixation) into energy (glucose) through photosynthesis, using the simplified chemical reaction:

Carbon dioxide + water + light energy – glucose + oxygen + water

In reality this process is far more complex than this equation would imply, it takes place through a series of steps, and the process of photosynthesis is not the same in all plants.

Three different biochemical methods for terrestrial plants to fix carbon have evolved according to variations in conditions. These are known as the C3, C4, and CAM pathways. C3 is the most common of these, whereas C4 and CAM are more recent evolutionary adaptations that have taken place in high-temperature/arid environments.

C3 pathway

Primarily cool season or temperate plants. Crop plants include many cereal grains (wheat, barley, rice, oats), spinach, beans, and cotton

The C3 pathway is the most common form of photosynthesis, and is the ancestral pathway for carbon fixation (Ehleringer and Cerling 2002). Still et al. (2003) estimated global coverage of C3 plants at 87.4 million km^2 or over 85 per cent of terrestrial plants.

In C3 plants, photosynthesis takes place in most leaf mesophyll cells (in chloroplasts), and the first stable product is a three-carbon molecule, 3-phosphoglyceric acid (hence the name C3 pathway). This cycle of reactions is often termed the Calvin cycle.

The C3 pathway is less efficient at high temperatures (which has implications should there be warming under climate change).

C4 pathway

Usually warm season or tropical plants. Crop plants include sugar cane, sorghum, and maize

In C4 plant photosynthesis, an extra stage takes place, where a four-carbon compound is produced as a first stage product before the C3 process (Calvin cycle) takes place. The C4 process probably evolved in grasses around 24–35 million years ago in response to carbon deficiencies, caused by factors such as reductions in global CO_2 or increases in temperature, drought, or salinity (Sage 2003). C4 plants are more efficient at utilizing nitrogen (N) and gathering CO_2 from the atmosphere and

Continued

Box 4.1 *(Continued)*

soil than C3 plants. This gives them improved carbon fixation where CO_2 or N is limited. However, in other conditions C3 plants are more efficient, as the less complex process uses less energy. Around 18 per cent of plants globally are thought to be C4 plants, almost of which are herbaceous (Still et al. 2003).

The protein content of C4 plants tends to be low compared to C3 plants (and this may have possible nutritional implications).

CAM pathway

Usually succulent plants, adapted to dry desert environments such as cacti. Crop plants include pineapple and agaves (which are often grown for fibre or for food products such as sweeteners)

Crassulacean acid metabolism (CAM) is a carbon assimilation pathway that has a high rate of water efficiency, and is a process of photosynthesis that evolved in plants suffering from water stress. CAM plants are able to store energy from the sun during the day and fix carbon overnight. By exchanging gases during the night, plant stomata can remain closed during the day, so reducing the rate of water loss through transpiration (see Chapter 6 for more details of transpiration).

Implications of different photosynthesis pathways for agroecosystems under changing climates

Most of the staple crops (including wheat, rice, barley, and oats) are C3 plants, which are less efficient under warmer, drier conditions or where water is limited/ unstable. Climate scenarios generally predict rises in both CO_2 and temperature. Although C3 plants will benefit under elevated CO_2 levels, this is likely to be offset by rising temperatures, which may even impact on the survival of these C3 crops. Overall yields of C3 plants such as wheat and other cereals are expected to decrease under predicted temperature rises, whereas C4 plants such as millet and sorghum yields may increase (Kumar et al. 2017).

C4 photosynthesis is more efficient under lower atmospheric CO_2 and/or higher temperatures (Ehleringer and Cerling 2002), and in the tropics, trees and shrubs are often replaced with C4 crops and grasses (Still et al. 2003). Unfortunately, although very productive, many C4 crops (with the exception of maize) are not really suitable for human consumption without processing, so they tend to be used more for energy crops or for animal food, and so are not considered energy efficient.

In order to try to deal with some of these issues, scientists are investigating ways to integrate C4 or CAM pathways into C3 crops in order to increase carbon fixation efficiency, or to help facilitate growth of more crop types under predicted stresses such as drought, salinity, or temperature increases (see Ming et al. (2015) and Davis et al. (2015) for example).

'mineralized' into inorganic forms in order for plants to access these nutrients. In some agricultural systems, inorganic nutrients are applied using chemically synthesized fertilizers, but they can also be applied as additional organic matter (in the form of manures and mulches, etc.). These in turn need to be mineralized, and there are numerous different organisms within agroecosystems, particularly in soils, that deliver these nutrient cycling services functions. Nitrogen cycling is described in Box 4.2.

Box 4.2 Nitrogen cycling in ecosystems: mineralization, nitrogen fixation, nitrification, and denitrification

Mineralization is the process by which organic compounds are converted from organic to inorganic forms and so become available to plants. Microorganisms but also animals and other organisms that decompose organic matter cause mineralization to take place.

Nitrogen (N) is required in the largest quantities of all the mineral nutrients for plant cells, with the herbaceous dry matter of plants typically containing 1.5–4.5 per cent N and protein containing around 12 per cent by weight (Forde and Clarkson 1999). Nitrogen comprises approximately 78 per cent of the atmosphere in the form of N_2 and is also contained in the soil within organic matter, but as neither of these forms is available to plants, both need to be converted into a form that is. Plants usually access N through their roots (the exceptions are insectivorous plants and those with N-fixing root nodules) as nitrate ($NO3^-$) or ammonium ($NH4^+$). Nitrate is the most readily available form for agricultural plants in temperate and subtropical regions, but this varies between soils and plant species (Forde and Clarkson 1999). Nitrate is more liable to leaching.

Nitrogen fixation is the conversion of atmospheric N into a form that plants can use. This can be either a natural biological process or carried out commercially to create fertilizers.

Organic N is converted to ammonium during the process of mineralization, but a further process called *nitrification*, is where ammonium is further converted by microorganisms to nitrite and then nitrate. Mineralization and nitrification both take place readily in warm, well-drained and well-aeriated soils.

Denitrification is a process that returns fixed N (i.e. nitrate) to its biologically inert form (N_2), and takes place under anaerobic conditions when microorganisms utilize the oxygen from the nitrate for themselves so the nitrate is no longer available for plants. It is common in saturated soils.

4.5.1.1 Nutrient cycling organisms

Much of the nutrient cycling in agroecosystems takes place in soil. Fungi (see Box 4.3) and bacteria are important carbon and nutrient cyclers, and nitrogen fixation in soils is largely carried out by them. Many nitrogen-fixing bacteria are free living, but others are mutualistic and have symbiotic relationships with other

Box 4.3 The role of fungi in agroecosystems

Fungi are single or multicellular eukaryotic organisms. They have membrane-bound nuclei and vegetative growth, but, unlike plants, do not have chlorophyll. They digest organic matter externally and then absorb it.

Fungi have a number of different biotic functions, which can be divided roughly into three functional groups based on how they obtain energy: saprophytic fungi, pathogenic fungi, and mycorrhizal fungi.

Saprophytic fungi

Saprophytic fungi are decomposers. They break down dead organic matter into fungal biomass, CO_2, and organic acids. This mineralization process releases nutrients that otherwise would not be plant available.

Pathogenic fungi

Pathogenic or parasitic fungi can colonize other organisms such as plant roots and are often considered agricultural pests as they can reduce plant productivity, but they can also be beneficial as they can be used to control other pests such as nematodes or other soil-living invertebrates Research into the role of fungi in supporting plant growth reveals ever more complex and surprising relationships. It has been shown, for example, that the entomopathogenic fungi such as *Beauveria bassiana* and *Metarhizium anisopliae* are not only common in the soil, but have an endophytic phase that confers insect resistance. These fungi provide a benefit to plants by killing their insect herbivores, and then translocate nitrogen from above-ground insect cadavers to the plant via fungal mycelia (Pineda et al. 2017).

Mycorrhizal fungi

Mycorrhizal fungi or mutualists have evolved to form symbiotic relationships with plant roots, where the fungi obtain photosynthesis-produced carbon compounds such as sugars from the plant, and the fungi help plants access both macro- and micronutrients such as phosphorus, potassium, nitrogen, copper, and iron; mycorrhizal fungi also produce a protein, glomalin, which helps bind soil particles together, so influencing movement of water, organic matter, nutrients, and soil microbes. This can support carbon storage.

Mycorrhizae can either be on the surface layers of the roots (ectomycorrhizae) or colonize the plant cells (endomycorrhizae or arbuscular mycorrhizal fungi). They may have a network of mycorrhizal strands, called mycelium, which can extend the plant roots, connecting plants to each other (see below).

Importance of hyphae and mycelium

An important part of multicellular fungi are threadlike filaments known as hyphae (singular hypha). Branching networks of these hyphae, known as mycelium, offer a variety of important functions in ecosystems, including improving soil structure by binding particles together. They can also connect plants underground through the

Continued

> **Box 4.3** (*Continued*)
>
> hyphal networks of mycorrhizal fungi. The hyphal connections between plants transfer water, carbon, nitrogen, other nutrients, defence signals, and allelochemicals between plants. These networks occur in all terrestrial systems including agroecosystems and are important for crop production as they can facilitate early plant survival, growth, and defence responses (Simard et al. 2012).
>
> It has been suggested that hyphae play a role in plant communication, transmitting infochemicals which aid plant protection by communicating systemic defence signalling across plant populations (Barto et al. 2012). Furthermore, hyphal networks are important in the redistribution of nutrients, so that plants growing in areas of high resource availability (e.g. high light or high nitrogen environments) transfer carbon or nutrients to plants located in poorer conditions (Simard et al. 2012). In agricultural systems, the impact of farming practice on these hyphal networks is poorly understood. However, there is some evidence that tillage and phosphorus application can adversely affect mycorrhizal networks by breaking hyphal links and lowering the extent of root colonization, respectively, but that increased crop diversity results in more diverse mycorrhizal fungi communities (Jansa et al. 2006).

organisms. These include rhizobia, which establish a symbiotic relationship with legumes (plants from the family Fabaceae which include peas, beans, and clover). Nodules are formed on the plant roots (or sometimes on the stems) and the bacteria colonize the nodules, where nitrogen fixation takes place. Legumes usually have high levels of plant protein (nitrogen is a key element in protein), but also high levels of nitrogen, which means they can be used within agricultural rotations to improve the nitrogen levels of soils for the next crop in the rotation.

Nematodes (or roundworms) have a variety of important functions in agricultural systems. Insect-parasitic nematodes can be important natural enemies of pest species, and plant-parasitic nematodes are often pests of agricultural crop species, but bacterial-feeding nematodes are the most abundant group of nematodes in agricultural soils and are another important group of nitrogen fixers. Proteins and other nitrogen-containing compounds from the bacteria they feed on are released as ammonium, increasing the amount in the soil that is available to plants. Nitrogen mineralization has also been found to occur at a high rate in soils where nematodes are present.

Earthworms carry out nutrient cycling activities, which can influence climate regulation and pollution remediation, although this is not always positive, with some evidence that earthworm activity can increase nitrous oxide (N_2O) (Blouin et al. 2013).

4.5.2 Soil engineering

A good soil structure is essential in agricultural ecosystems (see Chapter 5). This can be influenced by management of soils but also by natural biological processes.

Bacteria and fungi are important as soil engineers. Both can help bind soil particles together: bacteria through the production of 'sticky' polysaccharides and glycoproteins, and fungi through the mycelium.

One group of important soil ecosystem engineers are the earthworms, which play a role in a number of ways, including soil formation and maintenance of structure. Earthworms influence soil formation by accelerating weathering of minerals, transforming minerals, accelerating humus formation, and moving organic matter around the soil profile. They may also impact on erosion processes, although there is some dispute as to whether these impacts are positive or negative (Blouin et al. 2013). Earthworms influence soil structure by either compacting or loosening soil, depending on the species. This can affect bulk density. They also affect porosity of soil and aggregate distribution. But these are not the only ecosystem service functions that earthworms provide; the engineering functions they carry out can have secondary impacts. One such secondary ecosystem service is water regulation, which earthworm engineering influences by changing infiltration rates, water storage in soil, and (in temperate climates) creating surface roughness, which can affect run-off (Blouin et al. 2013).

Plants are also important soil engineers. Their roots help bind soils together and the ground cover they provide helps prevent erosion. In many arable agricultural systems, plants are removed regularly from the system, impacting on the soil structure and surface stability.

4.5.3 Water quality

Pesticides, together with nitrates, phosphates, and sedimentation, through run-off, erosion, and leaching to ground water, have reduced farmland water catchment quality.

Natural vegetation can provide buffers against movement of water and sediment into watercourses. Ground cover helps prevent surface flow, while roots slow subsurface flow and take up excess nutrients. Vegetation can also act as a physical barrier from pesticide spray drift (Haddaway et al. 2018). Cresswell et al. (2018) collated studies that looked at plant traits to deliver ecosystem services on farmland, and found 23 different plant traits that could deliver water protection. Although most of the studies were not specific to agroecosystems, traits such as root structure and density, and leaf area and biomass could help aggregate soils and slow the surface movement of water. Root structure (e.g. the number of fine or fibrous roots) and biomass were also found to influence take-up of potential water pollutants such as nitrates and phosphates (e.g. Sullivan et al. 2000; Read et al. 2010).

4.5.4 Primary production

Plant productivity has been shown to increase with plant species richness. This is partly due to the specific traits that different plant species offer. For example, Tilman et al. (1996) linked high plant productivity in species-rich grassland with efficient use of soil mineral nitrogen due to the diversity of species, but patterns are not consistent across systems and studies. Cardinale et al. (2011) carried out a series of meta-analyses examining the role of diversity for productivity. They found that, on average, declining diversity of plants leads to decreases in biomass of producers and may also reduce the rates of decomposition. They also surmised that these relationships are driven by some form of complementarity among species. However, they did not find any evidence that diverse polycultures outperform the most efficient producers within systems. Increased diversity can be of most value in regions where conditions are uncertain—for example, in the humid tropics where trees and shrubs reduce erosion and improve nutrient cycling. The phenology of agricultural products may also differ when grown under different conditions. For instance, Muschler (2001) found that Central American coffee yields were of better quality in agroforestry systems due to lower and more uniform ripening of the berries.

Plant growth-promoting rhizobacteria (PGPR) are found in the rhizosphere (the interface between plant roots and soil), or on or even in plant roots, and can enhance plant growth through symbiotic relationships with plants. Plant exudates provide nutrients for the bacteria, and in turn the bacteria provide a number of benefits for the plants, ranging from improved nutrient uptake through nitrogen fixation, to provision of compounds such as phytohormones, or by enhancing plant resistance to pathogens (Beneduzi et al. 2012).

In agricultural systems, earthworms are known to influence both yield and biomass. A meta-analysis by van Groenigen et al. (2014) showed that above-ground biomass was increased significantly with earthworms. The average increase was 31 per cent across major grain crops (based on 154 included studies) and 24 per cent in pasture grasses (106 studies). The effect in pastures was greater where there were no legumes. The researchers also found that the effects were greater overall in tropical/subtropical regions (58 studies) than in temperate/continental regions (113 studies), and were also greatest in clay soils.

4.5.5 Pest regulation

In natural systems, plants have a variety of strategies to avoid pests and diseases. Some, particularly trees, contain chemical toxins to reduce leaf digestibility, while others, such as annuals, rely on their ability to hide in time and space. Crop species are often conspicuous and lack chemical defences, so are much more vulnerable to pests and diseases.

One of the problems with chemical control is the development of resistance in target species. The most commonly cited early case of resistance was by Melander (1914). In 2008, a Michigan State University database contained records of over 550 insect pest species that had become resistant to at least one insecticide, and 7,747 cases of resistance relating to one or more of 331 different insecticide compounds (Whalon et al. 2008), and this resistance is continually developing. Neonicotinoids, a relatively new type of pesticide, were first used commercially in the mid-1980s, but by 2015 resistance had been found in over 20 insect pest species, including some significant crop pests such as *Bemisia tabaci, Myzus persicae, Aphis gossypii,* and *Nilaparvata lugens* (Bass et al. 2015).

Other aspects of modern agriculture can influence pest numbers; for example, simplified agricultural landscapes have been shown to have a 46 per cent lower level of pest control by insect natural enemies of crop pests (Rusch et al. 2016). Natural enemies can be defined as organisms that 'kill, decrease the reproductive potential of, or otherwise reduce the numbers of another organism' (Flint 1998).

There are three main types of natural enemy—predators, parasites, and parasitoids—and pathogens, although plant feeders, and even plants themselves may also be considered natural enemies of plant pests. Some key natural enemies in agricultural systems are outlined in Table 4.1.

Table 4.1 Key natural enemies in agroecosystems

Natural enemy	Definition	Characteristics	Agricultural examples	Methods of enhancement
Predator	Consumes more than one prey individual during its lifetime	Usually larger than prey species. Free living. May be 'specialist', feeding on only one (or closely related) species, or 'generalist/ polyphagous', feeding on a broad variety of species	Large invertebrates (e.g. beetles, spiders), small mammals, and birds that feed on smaller invertebrates or pest species	Conservation, augmentation, classical biological control
Parasite	Develops on or within host species	May or may not kill host, but usually reduces its fitness in some way	Parasitic fungi of agricultural pests. Nematodes of invertebrate pests	Usually augmentation or classical biological control
Parasitoid	Develops on or within a host during its immature lifestage	Always kills host. The adult form is usually free living	Parasitic wasps, flies, and occasionally beetles	Usually augmentation or classical biological control
Pathogen	A microorganism that lives on or in a host. It can be fungal, bacterial, or viral	Spread by contact, spores, etc.	Bacteria (e.g. *Bacillus thuringiensis* or Bt), viruses, and fungi	Augmentation or classical biological control. Also genetic engineering (especially for bt gene)
Plant feeders/ herbivores	Feed on weed plant species	May be vertebrate or invertebrate (plant-feeding insects are known as phytophagous insects)	Scale insects, sheep/goats	Conservation, augmentation, or classical biological control

The enhancement of these natural interactions to control pests can be termed bio-logical control. Biological control can be dated as far back as the beginning of the fourth century AD, when records show that citrus ants were sold at markets to mandarin orange growers to protect fruits from damage by other insects. The use of these ants is still ongoing, with bamboo 'corridors' between trees being used since at least the seventeenth century to facilitate dispersal of the ants (Huang and Yang 1987).

There are three main ways in which natural enemies in agroecosystems can be managed. These are importation or classical biological control, augmentation, and conservation.

4.5.5.1 Importation or classical biological control

Many of the pest species in agricultural systems are exotic (or non-native) to the system, and have been introduced alongside the crops and livestock that are their hosts. These species have often lost their own natural predators and pathogens, which can exacerbate their pest attributes. Importation (or classical biological control) is the sourcing and introduction of new (non-native) species into an (agricultural) environment in order for them to establish and control an existing (but often exotic) pest, and the method has been much used.

By the year 2010, over 2,000 different invertebrates (mainly predatory and parasit-oid insects) had been introduced into more than 200 countries to control insect pests. In addition, over 1,500 organisms to control weeds and nearly 150 patho-gens had been introduced outside of their native range (Hajek and Eilenberg 2018). This method has been very successful where the imported species has established successfully in its new environment, and where the pest and biological control agent are carefully matched. Classical biological control tends to work best in perennial systems, forestry, orchards, and other more permanent systems, as these offer better opportunities for the introduced species to survive (Hajek and Eilenberg 2018). However, there are many examples of classical biological control that have been considered failures as the control organism has failed to establish; but often where one biological control agent has failed, another has been used in its place more successfully. There are also examples of biological control agents that have themselves gone on to become pests, such as the cane toad (see Box 4.4).

4.5.5.2 Augmentation

Augmentation is the breeding and release of natural enemies to keep numbers of pests under control, where natural numbers are either too low for control or the natural enemy is not present. This strategy can be designed for short-term control of (predicted) pest outbreaks or utilize periodic release to allow for more sustained control of pests. Short-term control often does not require the natural enemy populations to reproduce, but instead biological control is managed in a similar way to chemical control agents, and the natural pest control organisms are peri-odically released when needed (e.g. at a vulnerable lifestage of the pest population,

---

Box 4.4 The cane toad

Cane toads (*Rhinella marina*) are native to the Americas, with a native range from Texas and into South America where they largely occupy rainforests and seasonal Amazonian savanna, although they are tolerant of a wide range of conditions. The cane toad was introduced into various regions around the Americas and Caribbean in the eighteenth and early twentieth centuries, and, after successfully controlling beetle infestations in Puerto Rico, were exported more widely to control invertebrate pests, but soon came to be considered pests themselves. They now occupy 47 countries as an invasive species.

Their diet is broad (they eat small terrestrial mammals, other amphibians, reptiles, birds, and insects, as well as plants and even domestic waste) and they exhibit a high tolerance to a wide range of environmental conditions. After introduction, cane toads quickly moved from agricultural into natural habitats, where they have impacted on native ecosystems through their voracious appetite and through an effective defence system, whereby they excrete toxins through the skin when under stress. In their new territories, few predators can tolerate the toxins and therefore there is no effective population regulation (Lever 2001). As a result, introduced cane toad populations have exploded; densities of toads in native areas have been shown to be just 1–3 per cent of that in the parts of Australia where they occur.

This has had two impacts on native species: (1) on local mammal and invertebrate populations through toads eating them, and (2) would-be predators were poisoned by the toads' toxins. The latter has had a well-documented and devastating effect on some native reptile populations; the toad has been responsible for the decline of monitor lizards in Guam and a wide range of predatory reptiles and land snakes in Australia. Efforts to control the cane toad are in development.

---

or when new pests move into the system). This process known as *inundative biological control* is often used in annual monocultures where the natural enemies of a specific pest are unlikely to occur in high enough numbers for stable populations. *Inoculative biological control* aims to provide longer-term biological control, through the release of organisms that have the ability to become self-sustaining (Hajek and Eilenberg 2018).

4.5.5.3 Conservation

Conservation is used to protect and enhance existing indigenous natural enemies through management and manipulation of the supporting environment through the provision of natural resources such as food and shelter. This can be in the form of semi-natural or uncropped areas within and around farmland, which provide a variety of resources, such as pollen and nectar for food, and varied structures for shelter (Shennan 2008). Increased plant diversity can improve the efficacy of natural pest control both through increasing the diversity of potential natural enemies and by supporting interactions in the system, so making it more stable (Gaba et al. 2014).

Natural enemies of weed species are often insects or other invertebrates, but may also be vertebrates. Deliberate introductions or management of natural enemies of problem plant species can have a number of advantages over chemical herbicides; once introduced, they may only need minimal effort to maintain, the cost can be lower than chemical methods, they are less disruptive to the ecology of the system, and do not require high levels of technology. That does not, of course, mean that they do not need to be managed carefully; the best natural enemy invertebrates for weed control are species specific, negatively impact both individuals and population dynamics of the target plant, are good colonizers, prolific, and will thrive within the host environment (according to Cornell University, Weed-feeders, https://biocontrol.entomology. cornell.edu/weed.php). As with all natural enemies, weed feeders should be introduced or encouraged with care so that they do not become pests themselves.

Plants can themselves provide pest control in systems through a variety of ways. The simple presence of plants can prevent invasive or pest plant species entering or dominating in systems as they do not have the space to move in, or their seeds are intercepted by other vegetation before hitting the ground. Most plants also contain chemicals that enable them to interact with other plants. These 'allelopathic' interactions can help control weeds, and some species secret weed growth-inhibiting compounds.

4.5.6 Pollination

Many plants cannot pollinate themselves (self-pollinate), or, if they can, the fruit produced is of lower quality. Pollination between different individuals (cross-pollination) needs a vector. This could be environmental, for example by releasing seeds to be carried by the wind or in watercourses, but in many cases is by an animal (most frequently, insects). Ollerton et al. (2011) estimated that between 78 per cent (temperate regions) and 94 per cent (tropical regions) of wild plant species are invertebrate/vertebrate pollinated, with more than 75 per cent of crops produced around the globe at least partially dependent upon insect pollinators (Klein et al. 2007), although this varies in different regions and temperate zones. Roubik (1995) estimated that about 70 per cent of tropical crops needed pollination at the time of his study. Common pollinators include invertebrates such as bees, birds such as hummingbirds, and even mammals such as bats.

Klein et al. (2007) used data from 200 countries and found that 87 leading global food crops were dependent upon animal pollination, with 28 crops not reliant. The latter include the four main staple grain crops, maize (or corn), rice, wheat, and barley. Thirty-five per cent of global production came from crops that depend on pollinators and 60 per cent from crops that do not depend on animal pollination (5 per cent were unevaluated). Of the animal-pollinated crops, the importance of pollinators varies, so for crops eaten directly by humans, which can be animal pollinated, these pollinators are essential for 13, production is highly pollinator dependent for 30, moderately for 27, slightly for 21, unimportant for 7, and of unknown significance for 9. Table 4.2 gives some examples of crops for

Table 4.2 Some common crops for which animal pollinator vectors are of greater or lesser importance for fertilization

Crop examples	Pollinators are essential or highly important	Pollinators are important or useful	Pollinators are not needed
Cereals			Maize, rice, wheat, barley
Nuts	Brazil nut, cashew		
Fruits	Kiwi, watermelon, passion fruit, pears, cherries, mangos, cucumber	Strawberries, broad beans, coconut, tomatoes, prickly pear	
Seeds/leaves/ roots	Buckwheat, turnip, coriander, fennel	Coffee, rape, sesame	

Source: Klein et al. 2007.

which animal pollination is essential, important, or unnecessary. For those plants for which animal pollination is partially dependent, pollination can impact on the scale of reproduction and on the quality of fruits produced, as illustrated in Figure 4.2, which shows self- and cross-pollinated raspberries (a) and blackberries (b). Poor fruit development of 'Chester' (a variety of blackberry) is particularly noticeable in fruits developed from bagged flowers that were picked on the same day from the same cane as unbagged ones (Figure 4.2b).

The economic value of pollinators is often to calculate, but Gallai et al. (2009), using a biometric approach, estimated the overall annual contribution of pollinators to agricultural crops to be around US $206 billion (or 153 billion Euros). More accurate estimates can be applied to specific food products and regions. Researchers at the University of Reading in the UK have been comparing efficiency of different pollinators on different crops in order to put a value on pollinators. Using this method, Garratt et al. (2014), for example, calculated that pollinating insects contributed a total of £36.7 million per annum to the production of two apple varieties in the UK. This was attributed to not only the quantity of production, but also the quality and size of apples, both of which impact on their market value.

Aizen and Harder (2009) showed that, overall, the area of flowering crops globally needing insect pollination is increasing, and at a faster rate than the number of bee colonies. In fact, many pollinator species are declining. Although the number of honey bee hives in Europe has increased since 1961, Potts et al. (2010) reported that in the USA there was a 59 per cent loss of colonies between 1947 and 2005 and a 25 per cent loss of colonies in central Europe between 1985 and 2005. Honeybees are threatened by multiple stressors including global increases of pests such as *Varroa destructor* (an ectoparasitic mite) and pathogens; environmental stressors such as malnutrition and exposure to agrochemicals; apicultural mismanagement; and lack of genetic diversity. These stressors interact, so that one will weaken the colony, resulting in an inability to resist another (Potts et al. 2010). There has also

Figure 4.2 Poor fruit development: self-pollination compared to cross-pollination in soft fruits. (a) A fully formed raspberry fruit (far left) that has been well pollinated is compared with fruit that is poorly formed (centre and right) due to inadequate pollination. (b) Comparison of fully formed blackberry fruit (top) with fruit from the same cane that has been prevented from animal-mediated cross-pollination by bagging (bottom). Photo: James H Cane

been increasing concern in Europe over the spread of the Asian hornet which arrived in France from China in 2004. Since then the hornet has spread across Europe. The hornet preys on beneficial insects, particularly the European honeybee *Apis mellifera* against which the bee has no defence. There is potential for the hornet to have a big impact on honey production in many countries given that honeybees are already under pressure from habitat fragmentation, lack of forage, and other pests and diseases. Efforts are being made to track invasions and to eradicate this serious pest (Budge et al. 2017).

Oilseed rape, a crop grown in several countries such as Canada, China, India, and Australia, and across Europe, has been identified as a crop particularly at risk of declines from bumblebees and honey bees as they are known to be the most effective pollinators of this crop (Stanley et al. 2013; FAO 2017). Furthermore, in California, watermelon crops have suffered a pollination deficit caused by declines in bee diversity due to reduced floral and nesting resources (Kremen et al. 2002).

Bumblebees and honey bees are not the only invertebrate pollinators that are declining. A large proportion of crop pollination services are carried out by wild bees and other insects such as hoverflies (Garibaldi et al. 2013), and these are also under threat. The potential costs of pollinator declines are often difficult to calculate, but in some regions there are indications that loss of pollination services are linked to yield declines. In India, for example, relative yields of pollinator-dependent vegetables appeared to either flatten or decline over an 18-year period. This trend was attributed to large-scale changes in pesticide use and in land management over the same period, and contrasted with non-pollinator-dependent crops that showed no such decline over the same period (Basu et al. 2011).

In some regions, pollinator declines are so extreme that natural pollination is no longer viable for productive purposes, leading to the introduction of non-native pollinators, hand pollination by people, or even diversification to products that do not need pollinating. One region where all three practices have been used to greater or lesser effect is Maoxian County in the Sichuan province of China. Traditionally, apple orchards have dominated the food production of this warm temperate valley, but multiple sprays of pesticides over four decades, combined with loss of natural habitats due to conversion to farmland, have led to a lack of natural pollinators. In some cases, this is exacerbated by a lack of suitable of pollinator trees within self-sterile orchards. The introduction of honey bees was promoted for a while by the local government, but this was largely unsuccessful, and those that did take up beekeeping lost many of their colonies due to the surrounding heavy pesticide use. Hand pollination of apples became the only option for farmers, and has been used on a large scale since the 1990s. But, this trend is, in this region at least, beginning to decline. The financial cost of employing hand pollinators, combined with decreasing productivity, and shifts in market demands have led many farmers in the region to diversify into other self-pollinated crops such as plums and walnuts, and many are also introducing intercropping of vegetables into these increasingly mixed orchards (Partap and Ya 2012).

4.6 Issues and challenges for management of functional groups and ecosystem services

There are a wide variety of challenges associated with the management of functional groups and ecosystems services in farmland, and these are often associated with the management practices that have developed over time to enhance yields of one or two target agricultural products. As discussed in Chapter 3, management of the land for a very limited number of organisms often directly conflicts with the management of the land for other organisms that may provide ecosystem services; for example, diverse vegetative areas can support valuable ecosystem services such as water quality protection, pollination, and natural pest regulation (Pfiffner and Wyss 2004; Reichenberger et al. 2007; Haaland et al. 2011), but to provide these more semi-natural areas on farmland, farmers have to reduce the amount of productive land.

Other methods of utilizing natural ecosystem services can conflict with contemporary intensive farming. The use of natural enemies for pest regulation is not designed to eradicate pests, but instead to reduce them to economically viable numbers. In contrast, pesticides usually aim to eradicate pests, but may lead to population explosions. In some regions, particularly in parts of the European Union and the USA, biological control is often undesirable, as farmers, retailers, and consumers expect to have completely clean crops, but in order for biological control to take place, and to sustain natural enemy populations, there needs to be a low level of the pest species in or around the crops.

Other impacts of modern farming can have unexpected impacts on the ecosystem functioning that takes place in agroecosystems. In 60 per cent of studies, herbivory increased under higher N (Shennan 2008), but of course increasing N in agricultural systems is often a key management measure in order to avoid crop deficiencies.

4.7 Tools for management of functional groups and ecosystem services

The commercial development of pest natural enemies, particularly microbial natural enemies (often termed biopesticides), is increasing in an attempt to replace chemical and potentially toxic synthetic chemical pesticides in agriculture. In 2016, the total world production of biopesticides was over 3,000 tons/year and this is increasing (Pawar et al. 2016). Some biopesticides use pathogens and viruses to control pest species; for example, they may employ insect pathogenic fungi. The success rate is variable, but although there have been inconsistent results in field trials using this approach, genetic modification of the pathogens is increasing the virulence and therefore the effectiveness and reliability of the method (Lovett and St. Leger 2018). Other biopesticides use a topical application of botanical bio-insecticides to kill pests; for example, cinnamon and vertiver essential oils have been shown to control sheep blowfly larvae (Khater et al. 2018). Although some biopesticides are biodegradable, others are more persistent and may not break down for months or years. This means that they require fewer applications, but there is potential for contamination in the wider agroecosystem, especially via water, which needs careful management (López-Serna et al. 2016). Biopest control is often used in conjunction with other natural and chemical methods, as illustrated in Box 4.5.

Similarly, commercial rearing of pollinators is increasingly utilized in many agro-ecosystems, and pollinator presence (both commercial and wild) can be encouraged

Box 4.5 A multifaceted approach to control of the black vine weevil

The black vine weevil (*Otiorhynchus sulcatus*) feeds on the roots and leaves of soft fruit and potted ornamental plants. Thought to have originated in Europe, *O. sulcatus* is now a pest throughout many temperate regions of the globe. Both the adult and larvae are pests, with the larvae feeding on roots and the adults on leaves. They reproduce parthenogenetically, and a single female can produce over 500 eggs, with pupae, adults, and larvae overwintering in the soil. Multiple methods of control of the black vine weevil have been developed. These range from cultural control and traditional biological control methods to biopesticides and chemical control, and target all life stages of the pest. Figure 4.3 shows the generalized lifecycle of the black vine weevil, with potential methods of biological and alternative control to target the different lifestages.

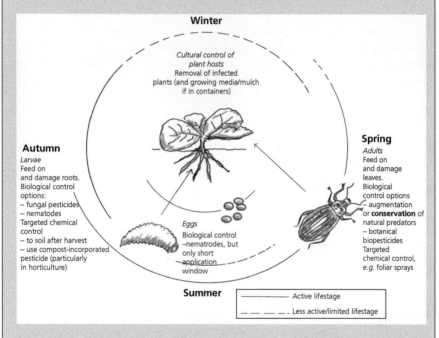

Figure 4.3 Simplified black vine weevil typical lifecycle in northern Europe, and potential integrated pest management (IPM) options. Knowledge of the biology and behaviour of target pest species and their hosts, and potential biological control agents can help support a successful IPM approach. Original line drawing: Nicola Randall

through the management of adjacent habitats. A study of commercial strawberries in Scotland in the UK showed that pollinator visits were increased by on average 25 per cent where there was a wildflower strip planted nearby, although it was unclear whether the most frequently observed visitor, the bumblebee, *Bombus terrestris*, was through greater retention of the commercial bee stock or wild bees that were attracted to the area (Feltham et al. 2015). In a similar study in American blueberry fields, although commercial pollinators (European honeybees) were similar in fields with and without wildflower strips, some of the native pollinators increased in the fields with wildflower strips. Over time, yields in the treated field also increased compared to the non-treated fields (Blaauw and Isaacs 2014).

Many other types of biotic function are being developed for commercial applications. The symbiotic relationships between plants and plant growth-promoting rhizobacteria (PGPR) can be exploited by researchers to improve yields and even nutritional content of crops. For example, plants can be inoculated with PGPR or treated with compounds that improve signals between the plants and the bacteria to enhance their growth, and this offers the potential to reduce the requirements for artificial fertilizers and chemicals (Backer et al. 2018).

Other, more traditional methods can make use of natural ecosystem functioning. Plant allelopathy has long been utilized by agricultural managers (e.g. see Weston 1996). Companion planting can be used in conjunction with other aspects of natural weed suppression provided by plants. For example, planting oats within an arable rotation in temperate regions can suppress weeds through allelopathy from their roots and through dense canopies (compared to wheat and barley). The pytotoxic compounds can also be utilized in mulches or even applied as plant extracts (e.g. see Głąb et al. 2017).

Figure 4.4 A farmer inspects her beehive India. Photo: Barbara Smith

Wider spatial structure is also important for the organisms that provide ecosystem services. More complex landscapes have been shown to provide refuges and corridors for organisms such as pollinators and natural enemies. A synthesis by Rusch et al. (2016) found that natural pest control declined by an average of 46 per cent in simplified landscapes, although a global synthesis of all arthropods by Lichtenberg et al. (2017) found that within-field diversity and other factors, such as the presence of organic farming, had a greater influence than wider landscape factors overall.

Socio-economic and policy initiatives can also drive ecosystem service provision. European farmers, for example, are required to buffer any waterbody next to arable land with a 2m-wide vegetative strip (European Commission 2018), as such strips are proven to mitigate pollution (Muscutt et al. 1993; Dorioz et al. 2006; Haukos et al. 2016). In India there are several small-scale initiatives training farmers to keep bees for both pollination services and honey production. Traditionally, subsistence farmers take honey from wild bees, but this is an unsustainable practice (farmers have reported wild bee declines (Smith et al. 2017)). In most cases, a training and support network is established to support farmers. Using locally built hives, the farmers are helped to introduce feral colonies of *Apis cerana*, often with the support of honey hunters who know how to locate colonies. See Figure 4.4. These schemes have been largely successful, increasing yields and producing honey as an output. *Apis cerana* is not as productive as the European *Apis mellifera*, but local co-ops enable farmers to sell the honey they produce (see https://utmt.in).

5
Soils

5.1 Introduction

Terrestrial food production is reliant on soil quality, yet astonishingly little attention is paid to soil management in agroecosystems. Soils were first officially recognized as a limited resource in 1935 when the USA passed its Soil Conservation Act. The Act was precipitated by what became known as the 'Dust Bowl' drought that decimated the Canadian and American prairies in the early 1930s. Many years of overgrazing were followed by deep ploughing of the delicate dryland soils to grow cereals and cotton, destroying soil structure and removing the deep-rooted vegetation which helped to retain moisture and physically bound the soil particles together. A period of relatively wet weather had masked the damage that was being done, but when drought conditions occurred, the unanchored soil blew away as dust in the typically high prevailing winds, affecting 100 million acres. The disaster led to poverty and displacement of rural people; the migration that followed remains the largest migration in American history in which 3.5 million people moved over a period of 10 years. It was an unprecedented disaster that prompted President Roosevelt to write that 'The history of every Nation is eventually written in the way in which it cares for its soil' and 'A nation that destroys its soil destroys itself'. Despite this stark illustration that soil protection is crucial, there has been little legislation governing soils in countries other than the USA. In Europe, a European Commission Soil Thematic Strategy was adopted in 2006 (EU Commission 2012), but only Germany and Switzerland have legislation that specifically targets the conservation of soil. In 2011 the Global Soils Panel (GSP) was established to focus on issuing guidelines for soil conservation, yet calls for international legislation were still being made in 2015 (Montanarella 2015), and by 2019 there was still no European Union legislation for soil as there was for both air (Air Quality Framework Directive) and water (Water Framework Directive), as a number of member states blocked its passage. There were guidance and good practice guidelines in a number of member states, however.

Not only does soil support food production, it also provides wider regulatory ecosystem services such as the capture and release of carbon, nutrients, and water, the detoxification of pollutants, the purification of water, and the suppression of soil-dwelling

The Biology of Agroecosystems. Nicola P. Randall and Barbara Smith, Oxford University Press (2020).
© Nicola P. Randall and Barbara Smith 2020. DOI: 10.1093/oso/9780198737520.001.0001

pests and pathogens (Smith et al. 2011). As the Dust Bowl disaster taught us, having a good understanding of soil ecology is key to managing soil sustainably; it is quite literally the medium in which our understanding of agriculture is rooted.

5.2 What is soil?

Traditionally, soil is considered as 'the natural medium for the growth of land plants… distinguishable from the initial material as a result of additions, losses, transfers, and transformations of energy' (United States Department of Agriculture (USDA) 1999). Soil is the product of weathered bedrock, the living organisms that interact intimately with it, and the products of the decay of those organisms.

Classically it is said that there are five soil-forming factors: climate, parent rock, topography, soil organisms, and time (Jenny 1994). Dokuchaev, the nineteenth-century soil scientist who identified these factors, only referred to natural processes and did not discuss the influence of human activity; however, since Neolithic times and the advent of farming, human activity has been a major influence on soil formation, conservation, and degradation. The rate of soil formation varies according to a variety of factors. For example, parent rock influences the rate of change depending on how easily it is weathered, and this is modified by climate, as temperature influences weathering processes: chemical weathering increases by two to three times with a temperature increase of 10°C. Similarly, in high temperatures, soil organisms will be more active and the rate of decay of organic matter increases, thus speeding up the accumulation of the organic component in developing soil. In contrast, nascent soil on steep slopes can be easily washed away, thereby slowing the rate at which soil develops (Jenny 1994).

Different members of the biotic community play varying but specific roles in soil formation; in some cases they may even be involved in primary weathering of parent rock. Bacteria and fungi are key players in this process. Fungi liberate minerals from parent rock through mineral tunneling, for instance, but animals can also play a role. An example can be found in the Negev Desert in Israel, where land snails feed on lichen that grows on limestone and in the process consume the substrate, thus breaking it down (Odling-Smee et al. 2003).

5.3 Types of soil in agroecosystems

At a very basic level, soil type can be distinguished primarily by the size and distribution of the particles and by the amount of organic matter. Soil texture is classified by the relative fractions of sand, silt, and clay particles. Classifications vary, but, approximately, sand particles are > 0.06 mm; silt particles are between 0.002 mm and 0.06 mm; and clay particles are < 0.002 mm. Using this method, soils can be categorized according to the predominant particle size (silty clay, sandy loam, etc.)

(Avery 1973). A loam is a soil with equal proportions of silt, clay, and sand. Soil texture has an impact on plant growth, nutrient availability, water retention, and soil biodiversity. In general, clay soils and those with large amounts of organic matter hold nutrients and water more readily than sandy soils. Sandy soils, which have large particle sizes and therefore larger pore sizes, are easily leached of nutrients, as water flows easily through them. Clay particles, which are small and flat, can pack tightly together, impeding water flow, in some cases to the extent that soils become waterlogged. Plate 4 shows the USDA soil texture classification. Soils are hugely variable and can be influenced locally in response to topography and local biota and at larger scales due to geology and climate.

Soil depth is also greatly variable, but sometimes difficult to establish. The upper limit can usually be identified as the boundary between soil and air, shallow water, plants, or still undecomposed plant materials (USDA 1999). The lower limit is often more difficult to establish, and may depend on whether there is a clear boundary between the soil and rock, or whether the signs of biological activity decrease more gradually (USDA 1999). Between the upper and lower limits of soil, characteristic 'layers' or horizons can be identified (Figure 5.1). The depth of the soil layers will

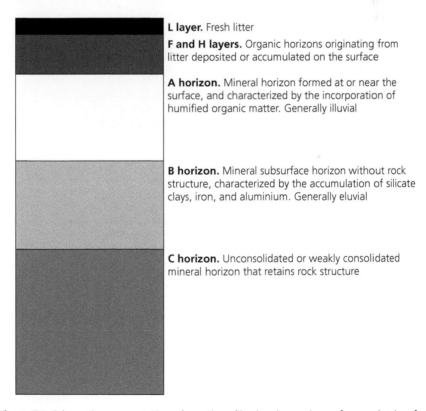

L layer. Fresh litter

F and H layers. Organic horizons originating from litter deposited or accumulated on the surface

A horizon. Mineral horizon formed at or near the surface, and characterized by the incorporation of humified organic matter. Generally illuvial

B horizon. Mineral subsurface horizon without rock structure, characterized by the accumulation of silicate clays, iron, and aluminium. Generally eluvial

C horizon. Unconsolidated or weakly consolidated mineral horizon that retains rock structure

Figure 5.1 Schematic representation of a soil profile showing major surface and subsurface horizons (Bardgett 2005).

also influence their ability to support different plants. Several different taxonomies of soils have been developed and there have also been various attempts to map the distribution of soils globally. (The FAO soils portal provides some links to some of these: see http://www.fao.org/soils-portal/soil-survey/soil-maps-and-databases/other-global-soil-maps-and-databases/en/.)

A typical example of a fertile soil is that of loess soils, which are predominantly aeolian (wind-blown) soils formed of very fine-grained quartz and lime; some of the fertility is associated with soil cation exchange capacity (i.e. the ability of these soils to retain plant-available, positively charged nutrients such as potassium and ammonium) related to the large surface area and negative charge of clay particles. In some loess regions the fine subsoil is overlain with high organic matter content. In the Eurasian Steppes and Canadian prairies of Manitoba, this results in black earth 'chernozems' which are the product of long-established grassland systems and are extremely fertile. As a consequence, these areas are often referred to as the 'bread baskets' of a particular region as they can, and do, produce cereal grain at scale. In contrast, in rainforests the nutrients are stored in the biomass rather than the soils, and the soils are generally sandy, fragile, shallow, and acidic with a high proportion of organic matter that is cycled quickly in the upper surface of the soil and characterized by a quick turnover of biomass. This can make them vulnerable to land-use change, such as agriculture. Without the litter from tree cover replenishing the organic component, the soils are quickly degraded and nutrients are leached from the system.

5.4 Importance and role of soil in agroecosystems

Soils are highly dynamic. Healthy soils provide a supportive growing structure for plants and regulate water availability (including controlling flooding and drought) and nutrient availability. They can also regulate climate through carbon and nitrogen storage.

Fertile soils are necessary to support agricultural production. A fertile soil is one that can supply sufficient macronutrients (such as nitrogen (N), phosphorus (P), and potassium (K)) and trace-elements (such as boron, chlorine, cobalt, copper, iron, manganese, molybdenum, sulphur, and zinc) to plants to enable growth. The minerals are derived from weathered bedrock, which may be the product of local bedrock or eroded bedrock transported by air (aeolian soils), water (fluvial soils), ice (glacial soils), or a combination of these events. Transported soils are frequently fertile as they contain material from a number of sources and consequently contain diverse minerals and nutrients. Nutrients are not only derived by the parent rock but also delivered by the living and decaying biological components of the soil. For example, N can be fixed from the atmosphere by some plants and delivered to the system via a complex relationship between plants and rhizobia. Rhizobia are bacteria that colonize root nodules in legumes and fix atmospheric N. There is increasing evidence of N fixation by non-rhizobia species too, such as bacterial endophytes (Puri et al. 2017). P is obtained from primary mineral sources and from mycorrhizal fungi, but also the decay of biological material (both plant and animal sources) as well as by direct atmospheric deposition. K is generally derived from decaying biological material such as animal manures and plant residues, but minerals such as feldspar are also an important source. It is made available through biological weathering, generally by fungi.

Organic matter is crucial for soil fertility, as it retains moisture and improves soil structure. It acts as a substrate for the soil biology too, with sufficient porosity to support the diverse soil organisms that support plant growth by regulating nutrient uptake, and can be considered the energy source that drives the biological system of the soil.

5.5 Soil structure

Soil structure is a key determinant of soil function, i.e. the extent to which it supports biotic life, and regulates carbon sequestration and water quality. Soil structure is expressed according to the stability of aggregates (how well particles stick together) and their size. A well-aggregated soil has space between the aggregates for roots to penetrate, water to flow, and air while a poorly aggregated soil impedes the movement of roots, air, and water (Bronick and Lal 2005). Good aggregation is essential for biotic organisms to thrive. Aggregation of particles is dependent on organic and inorganic soil colloids (small particles of humus and clay < 0.001 m, with a negative charge) and associated cations (positively charged molecules). Chemistry is important at the very small scale. The negatively charged soil colloids repel each other, but the presence of the positively charged cations facilitates the flocculation of particles (they act as glue to stick them together). Different cations are known to have different flocculating power: for example, sodium cations are poor flocculators, while calcium, magnesium, and hydrogen cations are good flocculators and promote aggregation. Under the right conditions, these microaggregates will stick to each other, resulting in macroaggregates, although they may subsequently disperse; aggregation and dispersal are influenced by soil chemistry, biotic material such as plant roots and fungal hyphae (which enmesh and release organic compounds that act as glue to hold particles together), and changing moisture levels and temperature. It is generally biology that determines small aggregates binding into bigger ones. One of the key players in this is glomalin, a glycoprotein excreted by arbuscular mycorrhizal fungi.

One measure of soil fertility linked to soil structure is cation exchange capacity (CEC) which is the total capacity of soil to retain exchangeable cations. This is important because many nutrients exist as cations (such as magnesium, potassium, and calcium), and the greater the capacity of the soil to hold onto these cations, the greater the amount of nutrients that are retained for plant growth. Although the cations bind to the negatively charged surfaces of the clay and organic particles, they retain a shell of water; this means they do not form direct chemical bonds with the surface, but form part of the diffuse surface above the soil particle and can be displaced by positively charged cations in the surrounding soil water, i.e. they are 'exchangeable' (Schaetzl and Thompson 2015). The cations in the soil water are available to plants. Clay soils with high levels of organic matter have a high CEC and are fertile, while sandy soils, which are comprised of chemically inactive quartz, have a low CEC and are inherently less fertile.

Soil management practices influence soil structure; management that minimizes soil disruption (minimum tillage for example) enhances aggregation and structural development and protects soil fertility (Bronick and Lal 2005).

5.6 Soil biodiversity

Soil biodiversity encompasses all living organisms within the soil: '*it is variation in soil life, from genes to communities, and the variation in soil habitats, from microaggregates to entire landscapes*' (FAO). In Box 5.1, soil biota are described according to size classes, and their respective function in relation to agriculture is discussed. Relatively little is known about soil organisms in comparison with their aboveground counterparts, but their function is increasingly being investigated; it is known that soil biota regulate important processes such as soil matter turnover and nutrient cycling, and drive crop growth yields. They also can produce or break down greenhouse gases—carbon dioxide (CO_2), methane (CH_4), and nitrous oxide (N_2O). While microbes are the primary drivers in C and N cycling, there is an indirect effect of biotic activity at higher trophic levels. Animals that consume microorganisms will influence rates of nutrient mineralization, and macrofauna such as earthworms can increase nutrient availability for plants (de Vries et al. 2013). The structure of the soil food web is illustrated in Figure 5.2.

Soil biota is often categorized by size as follows: microbes < 100 μm; microfauna < 100 μm; mesofauna 100 μm–2 mm; and macrofauna 2–20 mm.

5.6.1 Microbes

Microbes are organisms with body widths of < 100 μm. The microbial community comprises bacteria, fungi, archaea, and viruses, which perform a variety of functions for agricultural production including plant growth regulation, disease suppression, carbon and nutrient cycling, and decomposition of organic material.

The microbial community in soil is hugely diverse; collectively, this community of species is known as the microbiome. Interest in how microbiomes function and in the benefits they confer is growing. While one school of thought assumes that with such high levels of diversity there must also be high levels of functional redundancy

Box 5.1 Examples of soil microbial communities

Bacteria

Bacteria are simple single-celled prokaryotes (i.e. they have no membrane-bound nucleus). They usually require a film of water in which to live and to move through the soil. Some cannot move independently, but others have flagella or cilia, or will move in response to chemical stimuli. Soil bacteria are dominated by four phyla—Proteobacteria, Actinobacteria, Bacteriodes, and Firmicutes—which perform important ecosystem services in the soil including improving soil structure and soil aggregation, recycling of soil nutrients, and water recycling. Bacteria are known to excrete exopolysaccharides which form microaggregates in the soil by binding soil particles (Degens 1997); these aggregates are building blocks for improving soil structure, which in turn increases water infiltration and water-holding capacity of the soil. Bacteria are primary consumers in the soil and break down complex organic structures, facilitating nutrient availability through N fixation and P, K, and Fe mobilization (Rashid et al. 2016). Endophytic bacteria (those that live for at least part of their lifecycle in plant tissues) can boost plant nutrient uptake by secreting siderophores, iron chelating compounds that transport iron across cell walls, as well as by fixing N. In the legume family, roots are colonized by nitrogen-fixing soil bacteria belonging to the Rhizobiaceae. In low nitrogen conditions, the bacterium enters the roots, where it stimulates the production of nodules where the *Rhizobium* converts atmospheric nitrogen into a form available to the plant to support growth. Endophytic bacteria can also enable plants to tolerate cold and drought stress by strengthening cell walls, and prime plant responses systemically to resist pathogens or pests by synthesizing inhibitory metabolites (Liu et al. 2017). Most bacteria are particularly sensitive to the hydrogen ion concentration in the soil; hence the bacterial community assemblage is strongly driven by pH.

Archaea and viruses

The role of archaea and viruses in agricultural soils is poorly studied. Archaea are prokaryotes, allied to bacteria although it is thought that they diverged from bacteria a very long time ago and so are only very distantly related (e.g. as humans are to earthworms or trees). It is known that they produce methane and play a role in nitrification in the soil.

Viruses are very small organisms, consisting of DNA surrounded by a protein coat. Functionally they are known to regulate bacteria and plant pathogen populations and play a role in nitrification.

Fungi

Microscopic fungi are also important in soils. Some have extensive networks of hyphae (long filamentous structures), which can, in some habitats such as grassland, comprise a large proportion of the biomass in soil. These far-reaching hyphae can bridge pockets of moisture, enabling fungi to thrive in dry conditions where

Continued

Box 5.1 (*Continued*)

bacteria may be inactive and the translocation and distribution of nutrients across soil space. Mycorrhizal fungi are key players in agroecosystems. Over 90 per cent of terrestrial plants form relationships with these fungi and they facilitate below-ground communication between plants. Like bacteria, fungi are primary consumers in the soil, breaking down organic structures and decomposing organic matter. Also like bacteria, they can have a positive influence on soil structure. Fungi do this directly, by physically entrapping soil particles with the hyphal network and pushing them together, and by the exudation of mucilages and polysaccharides which bind particles in aggregates (Rashid et al. 2016). More information on the role of fungi in agroecosystems can be found in Box 4.3.

Foodweb pyramid in 1 m2 of soil square meter of soil

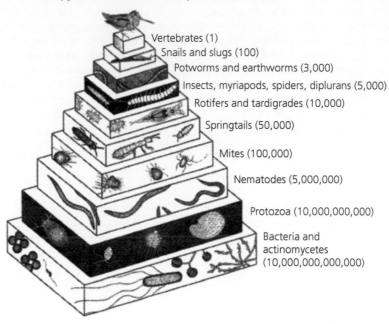

Vertebrates (1)
Snails and slugs (100)
Potworms and earthworms (3,000)
Insects, myriapods, spiders, diplurans (5,000)
Rotifers and tardigrades (10,000)
Springtails (50,000)
Mites (100,000)
Nematodes (5,000,000)
Protozoa (10,000,000,000)
Bacteria and actinomycetes (10,000,000,000,000)

Figure 5.2 A representation of the soil foodweb. Reproduced from James Nardi, 2007, The Life of Soil, with permission from the University of Chicago Press

(Grządziel 2017), there is increasing support for the idea that the thousands of taxa interacting in complex networks in soil provides multifunctionality in the higher plant community. The link between whole-soil microbiomes and above-ground plant and animal communities is emerging, opening up a new field for research. As an example, it has been shown that the abundance of *Aphis jacobaeae* (a specialist foliar-feeding aphid which lives on ragwort (*Senecio jacobaeae*)) was linked to the composition of soil microbial communities used by its host. The authors suggested that different soils hosted different fungal communities, which influenced the concentration of amino acids in the phloem sap, which in turn influenced the aphids (Pineda et al. 2017). Similarly, plant–soil microbe interactions can influence plant invasions. For example, black cherry (*Prunus serotina*) invasions in north-west Europe are thought to be in part due to their escape from the soil pathogens that reduce performance in their native range within the USA (Reinhart et al. 2003).

More information on soil microbes is provided in Box 5.1.

5.6.2 Microfauna

Microfauna are small animals with body size < 100 μm. In soil communities they are dominated by protozoa and nematode worms. Both need a film of water in which to survive and both have strategies for stasis to endure dry periods.

5.6.2.1 Protozoa

Protozoa are single-celled eukaryotes that readily form cysts and can be divided into four groups based on morphology: flagellates, ciliates, naked amoeba, and testate amoeba. They are largely bacterial feeders, and this process is key for soil fertility. When N is inside bacteria it is not available for plant uptake. Protozoa are responsible for cycling about one-third of this N and making it plant-available through grazing on bacteria. There are also protozoa that are predators, fungal feeders, and saprophytes.

5.6.2.2 Nematodes

Nematodes are round worms. They can be divided according to feeding habit, which can be identified by their mouthparts. They include plant feeders which feed on vascular plants, fungal feeders which feed by penetrating hyphae, bacterial feeders, animal predators, and omnivores (Yeates et al. 1993). Nematodes can play important beneficial functions in agriculture, primarily in soil fertility and in biological control; however, they are also plant parasites which feed on plant roots. A small number of these species can do high economic damage; however, as nematodes are highly host specific, where it is possible to introduce well-designed rotations the impact can be minimized. In potato production there are persistent problems with nematodes *Globodera rostochiensis* and *G. pallida*. *Globodera pallida* in particular has become widespread—in part due to its prolonged hatching period but also because varieties that are resistant to *G. rostochiensis* but not *G. pallida* have been widely grown. Nematode control is optimally managed using an integrated approach of variety choice, sustainable rotations, and judicious use of nematicides.

Both nematodes and protozoa play a beneficial role in making nutrients available to plants. These organisms can excrete N in the form of plant-available inorganic ammonium compounds through feeding on organic material. By grazing on bacteria and fungi, they break down organic matter and redistribute nutrients in the system and also play a role in pest and disease control by feeding on pathogenic bacteria.

5.6.3 Mesofauna

Mesofauna are 100 μm to 2 mm in size. They mostly comprise microarthropods, including Collembola (springtails), Acari (mites), and Enchytraeids (potworms). Their main role is as detritivores as they accelerate the turnover of plant matter to soil organic matter (SOM) and so increase mineralization and nutrient availability. Some Acari are also important plant pests (see Box 5.2).

Box 5.2 Examples of mesofauna

Collembola

Collembola are the most abundant of soil mesofauna. The majority are < 6 mm and have up to six abdominal segments. They can be epigeic (surface dwelling) hemiedaphic (litter dwelling), or edaphic (subsurface dwelling).The epigeic species generally have a small tail-like appendage known as a *furcula* folded beneath the abdomen which they use to jump when threatened. Collembola can be found in the majority of the world's terrestrial ecosystems, living in both wet and dry habitats in high densities. They can be found in abundances of up to several million individuals per square metre, with hundreds of different species found within some local ecosystems. These numbers can be impacted by soil management such as intensive agriculture, however (Rusek 1998). Collembola are usually highly specialized feeders. Some species feed on microbiota such as bacteria and fungi and have strong species preferences, through which they regulate bacterial and fungal populations. Others degrade and shred dead organic matter and play a role in litter breakdown, augmented by those species that feed on the feces of other soil animals, breaking it down as part of the process that forms humus. Their droppings also form an important part of the soil matrix.

Acari

Acari (mites) are small hard-bodied invertebrates with eight legs, allied to the arachnids. Their life history is diverse and Acari include parasites, predators, and detritivores (Gulvik 2007). Several species are known to be a pest, particularly the polyphagous red spider mites (*Tetranychus* spp.) which feed on a wide range of fruit and vegetable crops and cause serious problems in both glasshouse and field conditions. For example, *T. evansi* is known to cause up to 90 per cent yield losses in tomato production (Fernandes et al. 2015) and *T. urticae* is a serious pest of both okra (Shukla 2018) and aubergine (Kumar et al. 2018). Concomitantly, predatory mites are used in pest control programmes. A recent success story is that of *Amblyseius swirskii* which was first introduced commercially in 2005 and has since been released in fifty countries. Its success is based on its ability to control a number of crop pests and its life history. *Amblyseius swirskii* can reproduce by feeding on pollen so that it's possible to build up a population of predators ready for pest outbreaks. In Spain, biocontrol in sweet peppers grown in protected cropping systems

Continued

Box 5.2 (*Continued*)

(glasshouses and polytunnels) increased from 5 per cent in 2005 when the mite was made available to almost 100 per cent in 2009—representing the protection of an area of 6,000 ha (Calvo et al. 2015).

Potworms

Potworms are members of the Enchytraeidae family. Although most studies on potworms have been carried out in forest systems, they can also be present in large numbers in arable fields and pastures; in temperate systems, 90 per cent of the organic matter in pastures is processed by earthworms and potworms. They are smaller than earthworms but more abundant, and carry out a similar functional role in soil food webs by feeding on bacteria, fungi, protists, hyphae, and decomposing organic matter. Living in the first 10 cm of the soil, they are influential in decomposing litter, in nutrient cycling, and in the fate of organic matter, and they influence soil structure by boring through the soil and depositing fecal pellets. They themselves are predated by nematodes, beetles, and centipedes (Pelosi and Römbke 2017). It has also been reported that potworms can control *Fusarium* plant pathogens and mycotoxins from plant residues, thereby protecting crops at risk from these factors (Oldenburg et al. 2008).

5.6.4 Macrofauna

Macrofauna (2–20 mm in size) are the most studied of the soil biota and come in a range of functional groups, including detritivores such as millipedes and woodlice, which break down organic matter, and centipedes and scorpions, which are predatory. Larvae of various other macrofaunal species live in the soil and may be pest species, such as Tipulid (cranefly larvae) and chafer beetle larvae which feed on plant roots or influence the nutrient cycle by breaking down organic matter in the soil. Five taxa that play an important role in agroecosystems are discussed in Box 5.3.

Box 5.3 Examples of macrofauna

Earthworms

Earthworms are a key indicator of soil quality. They are known to be important for improving and maintaining soil fertility, soil structure, and aggregate stability (Riley et al. 2008). They are considered to be ecosystem engineers due to their role in chemical and mechanical weathering, the creation of humus, and biomass turnover (Blouin et al. 2013). The function of earthworms is explored more fully in Chapter 4.

Continued

Box 5.3 *(Continued)*

Termites

Termites are eusocial invertebrates in the order Isoptera. They live in dense colonies with a single queen and exhibit a range of functional feeding types, including those that feed on woody matter, fresh herbaceous material, fungal material and lichens, decomposing vegetative material, and soil (Black and Okwakol 1997). Nests can be in excavated galleries below ground or in wood, protruding above ground (in characteristic mounds) or in trees, although they always have protective shelter tubes connecting them to the ground (Noirot and Darlington 2000). They are well known as agricultural pests in the tropics. Their presence and abundance are strongly affected by management in agricultural systems and varies between cropping types (Black and Okwakol 1997). Mechanically, termite activity has a positive influence on soil hydraulic conductivity and infiltration rates. Termites also play a functional role in nutrient cycling and nutrient redistribution: mounds are known to be sites of high microbial diversity and activity, and they are known to fix N with gut bacteria and provide nutrients in the soil. Through these routes termites support plant health (Holt and Lepage 2000), but they are also a significant pest of a number of crops including trees in plantations and orchards, coconuts, palms, sugar cane, rice, maize, wheat, sorghum, groundnuts, coffee, tea, cocoa, yam, cassava, and cotton (Rouland-Lefèvre 2010).

Ants

Like termites, much of ant influence on soil is due to the mechanical effect of nest construction, galleries, and mounds. Ants modify soil profiles through transporting and rearranging soil particles, and these activities influence bioturbation, soil structure, and water infiltration, although the impacts on soil health can be both positive and negative. For example, in some cases improved porosity increases infiltration, but in others soil particles can be distributed so that there is surface compaction which can lead to water run-off and erosion. Ants bring dead plant and animal matter to their nests, thereby redistributing nutrients through the system, and it is thought they increase carbon and nutrient levels, especially N, P, and K, as well as exchangeable Mg and Ca (De Bruyn and Conacher 1990).

The majority of ants are omnivores, combining predation with scavenging and feeding on plant-based foods, although some species specialize on plants or fungus or are hunters. Many taxa supplement their diet with honeydew or nectar (Blüthgen and Feldhaar 2010). Like termites, some are able to fix N due to symbiosis with N-fixing bacteria (Pinto-Tomás et al. 2009). Ants form mutualistic relationships with insects that secrete honeydew including aphids and Lycaenid butterflies (Blüthgen and Feldhaar 2010).

Continued

Box 5.3 (*Continued*)

Spiders

Spiders are hard-bodied invertebrates with eight joint legs. They are predatory species belonging to the Araneae, which is the largest order in the class of Arachnids with 30,000 species over sixty families. In the agroecosystem they play a useful role as pest control agents, as they are obligate carnivores and do not damage plants, while consuming a wide range of prey. Spiders have different hunting modes, being either web weavers or hunters (Rajeswaran et al. 2005). Webs can be of several types including funnel webs, orb webs, and sheet webs. Spiders can have an indirect effect on primary productivity and elemental cycling through trophic cascades (Liu et al. 2015). In this context, a trophic cascade describes the indirect effect of a change in predator behaviour where prey is no longer controlled and leads to an impact on organisms, communities, and processes further down the food chain. For example, it has been shown that changes in spider activity can impact litter decomposition (Liu et al. 2015). In another example, grasshoppers, stressed by the presence of spiders, change their diet to one that is more carbohydrate rich which has consequences for the amount of N released when grasshopper carcasses break down. The reduction in available N reduces the rates of decomposition of vegetative material by microbes and therefore the nutrients available for plant growth (Hawlena et al. 2012). Although spiders do not directly impact soil processes, they do so via top-down regulation.

Beetles

The Coleoptera are the most diverse order of insects, comprising 400,000 species. They are distinguished from other insects by their elytra (hard wing cases that are adapted forewings). In soil, beetles apply top-down regulation as predator species and also have an impact on nutrient cycling as many species use dead woody material and vegetation as a food source. Dung beetles (Scarabidae) are particularly important in soil processes. They are globally distributed but have their highest diversity in tropical forests and savannas. Dung beetles are largely coprophagous but also feed on rotting fruit, carrion, and fungus and can be categorized into three ecologically functional types: paracoprid (tunneler) species bury dung in tunnels close to the original site of the dung; telocoprid (roller) species roll the dung into a ball and move it some distance away before burying it below the surface; and endocoprid (dweller) species brood their young inside dung (Nichols et al. 2008). Dung beetles play a number of roles in agricultural soils including nutrient cycling, bioturbation, plant growth enhancement, and secondary seed dispersal. In nutrient cycling, dung beetles physically relocate nutrients by moving dung and accelerate the mineralization of N by regulating microbial communities (Nichols et al. 2008). Some researchers have reported higher nutrient levels in soils exposed to dung beetles (P, K, N, Ca, and Mg) (Bertone 2004).

5.7 Issues and challenges for soil management in agroecosystems

All soils are degraded over time and older soils will be less fertile. The FAO defines degradation as 'a change in the soil health status resulting in a diminished capacity of the ecosystem to provide goods and services for its beneficiaries'.

Anthropogenic soil degradation can be defined as 'a process which lowers the current and/or future capacity of the soils to produce goods or services' (Oldeman 1992). Figure 5.3 shows processes through which soil degradation can affect the environment.

Oldeman (1992) recognized two categories of soil degradation processes: those relating to the displacement (or erosion) of soil material, and those relating to in-situ deterioration. He attributed water and wind forces as the main drivers for displacement of soil. Many agricultural practices can contribute to soil degradation; in arable systems, soils can be left exposed with no vegetation following cropping, and in temperate regions this often coincides with times of the year with increased precipitation leading to erosion of the topsoil; overgrazing reduces vegetation and exposes soils; heavy machinery or high livestock numbers can compact soils; and pesticides can impact non-target beneficial soil biota. Soil degradation is generally characterized in a number of ways including desertification, salinization, erosion, compaction, or encroachment of invasive species (Gibbs and Salmon 2015). Table 5.1 summarizes some different types of soil degradation and illustrates potential agricultural causes and impacts. The principal causes of soil degradation vary regionally; Figure 5.4 illustrates some of the known regional variations.

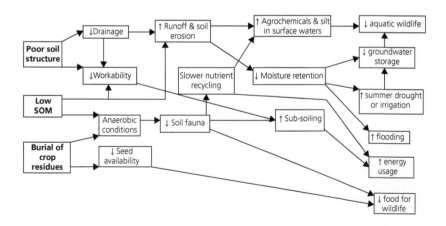

Figure 5.3 Routes through which degraded soil affects the environment (from Holland 2004).

Table 5.1 Different types of soil degradation, agricultural practices that can lead to them, and their potential impacts on agricultural production

Category (in bold) and type of soil degradation	Potential causes from agriculture	Characterization of degradation	Biological implications in and around agricultural systems
Displacement of soil			
Water erosion	Compaction or crusting of soil leading to reduced infiltration capacity	Blanket loss of (usually fertile) topsoil	Reduces production capacity (yield) due to loss of nutrients and structural changes to surface structure, influencing microclimates. Surface run-off changes can impact on the plants and animals within that piece of land, but also impact on surrounding aquatic ecosystems by increasing pollution from sediments and nutrients
	Farming on steep slopes	Terrain deformation (gullying)	
Wind erosion	Decreases in vegetation cover (e.g. due to overgrazing or crop removal)	Uniform loss of topsoil	Reduces production capacity (yield) due to loss of nutrients and soil structure. Sediments often are lost to aquatic ecosystems, where they can cause pollution
		Terrain deformation (dunes and hollows)	Changes to surface structure influencing microclimates, and may damage crops and other vegetation. Can get into aquatic ecosystems
		Overblowing (of soil particles into other areas)	
In-situ deterioration			
Chemical degradation	Intensive/long-term agriculture on poor-fertility soils, lack of manure or fertilizer application, clearing of natural vegetation, irrigation, groundwater/seawater intrusion due to water diversion, excessive groundwater use	Loss of nutrients and/or organic matter	Nutrient availability directly impacts soil biota and plant health. Organic matter influences soil structure, nutrient availability, and water-holding capacity, all of which directly impact soil organisms and plants
		Salinization	Impacts on soil biological activities and on plant health including productivity (yield)

(Continued)

Table 5.1 *Continued*

In-situ deterioration

	Drainage or oxidization of pyrite-containing soils, overapplication of acidifying fertilizers	Acidification	Reduces agricultural potential by increasing leaching of key nutrients. It may also lead to toxicity in soils, and to livestock health problems due to low levels of leached nutrients such as calcium
	Excessive pesticide use, excessive manure application	Pollution	Toxicity to plants and livestock, pollution of surrounding aquatic ecosystems
Physical degradation	Heavy machinery	Compaction	Structural changes to soil impact on soil biota and plants, for example by influencing water availability, root penetration, and microbiota activity.
	Reductions to organic matter combined with insufficient protection from precipitation to exposed soil	Crusting and sealing	Limited oxygen and water flow into and out of soils. Restricts seedling emergence and plant growth and influences soil biota activity
	Drainage interventions leading to flooding elsewhere	Waterlogging	Low levels of oxygen in plant root zones impact on seedling emergence and development and on later plant growth. May cause loss of some deeper roots
	Drainage and/or oxidization	Subsidence	Crop yield impacts due to changes to soil structure and water distribution

Source: Adapted from Oldeman 1992.

5.7.1 Soil erosion

Total soil erosion is defined as 'the total amount of soil lost by all recognized erosion types'. In many cases, soil is eroded faster than the natural replenishment rate; for example the USA is losing soil ten times faster and China and India are losing soil thirty to forty times faster than the rate at which it is replenished naturally (Pimentel 2006). In Europe, on average, actual soil erosion rates are three to forty times greater than upper limits for 'tolerable soil erosion', although there is significant

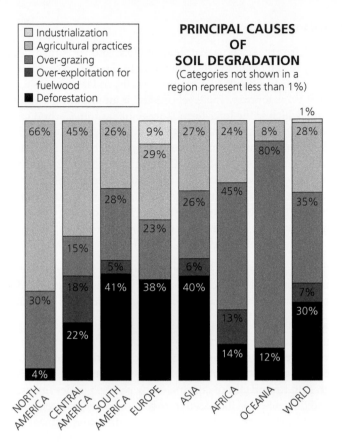

Figure 5.4 Principal causes of land degradation in different regions of the world. source: https://www.fewresources.org/soil-science-and-society-were-running-out-of-dirt.html

spatio-temporal variation. Tolerable soil erosion can be defined as 'any actual soil erosion rate at which a deterioration or loss of one or more soil functions does not occur' (Verheijen et al. 2009).

As a result of erosion, 30 per cent of the world's arable land has become unproductive in just 40 years. Approximately 60 per cent of soil that is washed away ends up in rivers, streams, and lakes, and this increases the risk of flooding and intensifying water contamination from fertilizers and pesticide run-off which has an impact on the biology of water courses (see Chapter 6). Globally, approximately 10 million ha of cropland is lost due to soil erosion each year (Pimentel 2006). Soils are increasingly degraded due to anthropogenic activity, much of it related to agriculture; land suitable for agricultural production is shrinking in many parts of the world (Gibbs and Salmon 2015). Many agricultural practices can accelerate the naturally occurring process of mobilization and deposition of soil particles, for example erosion rates in ploughed fields are ten to a hundred times greater than the rate of natural soil formation (Cameron et al. 2015).

5.7.2 Soil quality

In addition there are concerns over loss of SOM and soil organic carbon (SOC), which are key to soil quality. SOC refers only to the carbon component of organic compounds. SOC is difficult to measure directly, so laboratories tend to measure and report SOM as a proxy. The level of SOC is an important indicator of *carbon sequestration*, which is the long-term storage of carbon. Carbon can be stored in soils as well as environments such as the oceans or vegetation. Although most of the Earth's carbon is stored in the oceans, nearly 80 per cent of the carbon in terrestrial ecosystems is stored in soils (Ontl and Schulte 2012), being stored in the organic matter component, and in the mineral component of calcareous soils. The length of time that carbon is stored is dependent on soil management. When carbon is released, a proportion is released as carbon dioxide into the atmosphere, contributing to climate change. In England and Wales it is estimated that soil is losing carbon at a rate of 0.6 per cent a year (Hirsch et al. 2016).

Poor management of soils can lead to the release of nitrogen oxides and dioxides (NO_x) which also contribute to climate change. It has been estimated that almost half the NOx emissions in California come from agricultural soils (Almaraz et al. 2018).

The degradation of soil has implications for the soil biota which govern carbon sequestration and essential nutrient cycling. Agricultural intensification has been shown to reduce SOC and the diversity and abundance of soil biota (de Vries et al. 2013), bringing about shifts in community assemblage. Microbial communities tend to shift towards those dominated by bacteria under intensively managed systems, and this can cause increased nitrogen losses and reduced carbon sequestration, whereas fungal-dominated microbial communities, typical of more extensively managed systems, are linked to conservative nutrient cycling and greater carbon storage (de Vries et al. 2013).

Breeding has created elite crops that are optimized for high nutrient inputs and that are dependent on chemical controls for both pests and diseases, disrupting the relationship between crops and microbes which extract nutrients from the soil and deliver plant protection. Repeated deep ploughing has had a negative effect on SOC, which in turn has led to a loss of soil structure and along with it water-holding abilities and nutrient supply as well as rapid erosion (Cameron et al. 2015). Industrial livestock production also has negative consequences for soil, particularly compaction, which itself causes a variety of challenges (see Box 5.4).

5.7.3 Salinization and desertification

5.7.3.1 Salinization

Salinization of soil is a critical problem in arid areas, as it is a precursor for desertification and leads to poor productivity and in extreme cases the abandonment of agricultural land (Amezketa 2006). The principal anthropogenic driver for salinization is using poor-quality water for irrigation (such as groundwater

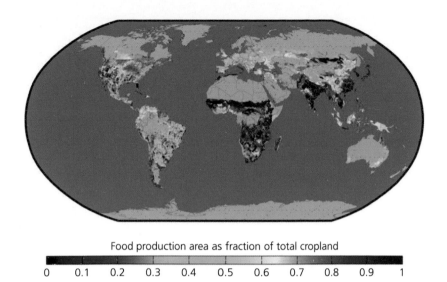

Food production area as fraction of total cropland

0   0.1   0.2   0.3   0.4   0.5   0.6   0.7   0.8   0.9   1

Plate 1 Food production area as a fraction of total cropland (Reproduced from Foley et al. 2011 with permission from Springer Nature).

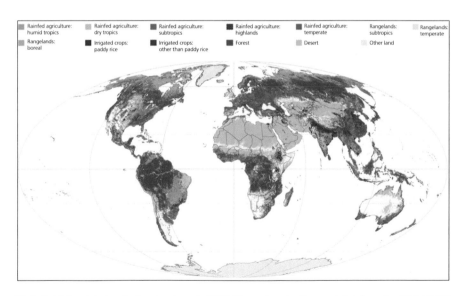

| Rainfed agriculture: humid tropics | Rainfed agriculture: dry tropics | Rainfed agriculture: subtropics | Rainfed agriculture: highlands | Rainfed agriculture: temperate | Rangelands: subtropics | Rangelands: temperate |
| Rangelands: boreal | Irrigated crops: paddy rice | Irrigated crops: other than paddy rice | Forest | Desert | Other land | |

Plate 2 Major global agricultural systems in 2011 (Reproduced with permission from the FAO, 2011).

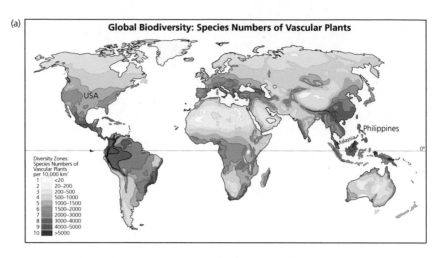

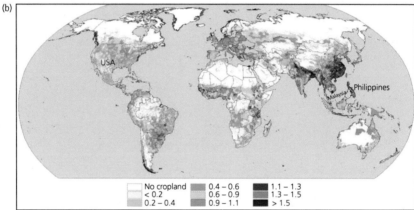

Plate 3 Maps of biodiversity and farmland intensity. (a) shows a world map of species richness of vascular plants (Reproduced from W. Barthlott et al, Erdkunde 61(4), 2007). (b) shows global patterns of crop use intensity (Reproduced from Siebert et al. 2010).

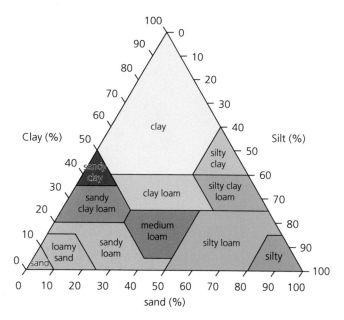

Plate 4 Soil texture classification as defined by the United States Department of Agriculture (USDA). (Source: Wikimedia, https://commons.wikimedia.org/wiki/File:SoilComposition.png.)

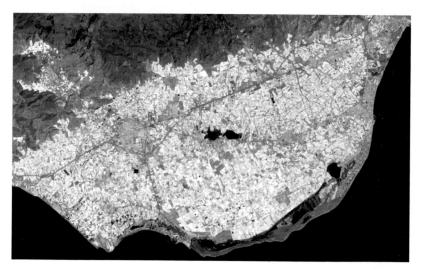

Plate 5 Glasshouses in Almeria province in southern Spain. The image covers an area of 19 × 30.5 km, was acquired on 20 July 2008, and is located at 36.75 degrees north latitude, 2.75 degrees west longitude. The glasshouses (which appear white on the image) cover almost the whole region. (Source: NASA/GSFC/METI/ERSDAC/JAROS, and U.S./Japan ASTER Science Team)

(a)

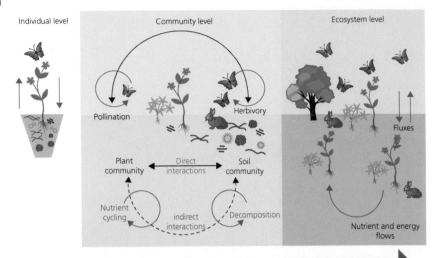

Increasing system complexity

(b)

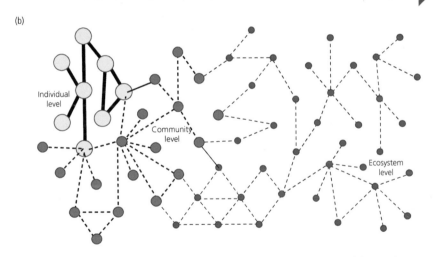

Plate 6 (a) There are three levels at which above-ground–below-ground interactions can be investigated. At the individual level, the focus is on interactions between few organisms in a controlled study system; at the community level, studies focus on the community members and interactions in the environment; at the ecosystem level, fluxes and processes are investigated at the broad field level. As the system increases in complexity, so the ability to determine mechanisms decreases and it is difficult to scale between the three levels of research. (b) Network analyses have the potential to link the many types of data across levels, resulting in nested knowledge relating individuals through community and ecosystems. (Reproduced from Trends in plant science, Vol 23 (9), Ramirez et al). Network Analyses Can Advance Above-Belowground Ecology. pp759–768, Copyright (2018), with permission from Elsevier

Box 5.4 Soil compaction

Globally, many soils have become degraded as a result of agricultural practices. A particular challenge is soil compaction, where soil structure deteriorates as a result of pressure from livestock trampling, farm machinery, or other farm vehicles, particularly when the soil is moist and soft (see Figure 5.5). Soils with a high clay content in regions of high rainfall are particularly vulnerable, but compaction also occurs on other soils such as on peat and sands with a high silt content.

Soil compaction can be a widespread problem, particularly during wet seasons, and is a real risk with the use of large tractors and heavy machinery capable of travelling in all conditions. When soil is compacted, soil porosity, hydrology, and air content are affected and soil biota is altered. For example, earthworms, a key organism for maintaining soil health, become scarcer—in highly compacted soils few earthworms will be found. Plant roots find it harder to penetrate compacted soil. When semi-natural permanent grassland becomes compacted, plant communities change as soil may become waterlogged at the surface, dry below the compaction zone, and harder for plant roots to penetrate. Grass and annual crop growth are noticeably poorer on compacted soils.

Soil compaction also leads to other major environmental problems associated with enhanced water run–off, which leads to soil and nutrients entering fresh and coastal waters, to the detriment of aquatic life. Run-off from compacted land also increases flood risk owing to the increased volume and speed of water coming off farmland. Within arable systems, compaction is associated with tillage (cultivation) and occurs in all types of tillage systems, including no-till, minimum till, and plough-based systems and occurs in both conventional and organic systems. Other operations

Figure 5.5 Soil compaction. Farm vehicles have accessed this land while wet, causing compaction and facilitating run-off on a soil that is inherently well drained. (Photo: Sue Everett.)

Continued

Box 5.4 (*Continued*)

such as spraying and harvesting cause localized compaction, especially from wheels. Turning areas (headlands) and tramlines are particularly affected.

Addressing soil compaction requires an understanding of the underlying soil type and taking actions such as modifying husbandry to break up the compacted layer (which may be near the surface or in the subsoil), adapting machinery, and ensuring future operations are appropriate and timely (e.g. not carried out when the soil is wet). In grassland systems, untimely grazing and overstocking will need to be avoided; attention must also be paid to the location of water troughs and supplementary feeding as animals tend to congregate in these areas which then lose their vegetation and become trampled and compacted.

contaminated by sea water), which directly adds salt to the soil. However, there are other routes including poor irrigation practices coupled with poor drainage conditions; irrigation in arid conditions with adverse evapotranspiration rates; contamination of groundwater due to unlined canals, reservoirs, or waterlogging; land clearing and poor drainage, all of which may mobilize salts so that they accumulate in the upper layers of the soil and reach the root zone where they are toxic to plants. In addition, waste waters, especially from industrial or mining operations, are often rich in salts and may be improperly disposed of or used for irrigation (Daliakopoulos et al. 2016).

Soil salinization has a negative effect on soil function and on aggregate stability. This reduces biodiversity and microorganism diversity, which can eventually lead to the elimination of all vegetation. A negative feedback loop of SOC is entered into, where reduced microbial activity drives fertility loss and less biomass production, which affects soil aggregation and promotes a higher level of plant matter in accumulating organic matter. This increases erosion rates. The impact of climate change is likely to exacerbate problems of increasing soil salinity (Daliakopoulos et al. 2016). Good irrigation practices are essential to mitigate soil salinization. Globally, 950 million ha in a hundred countries are salt affected, mostly in arid regions of Africa, Latin America, some US states, Asia, and Australia. This equates to approximately 10 per cent of the total land surface (Schofield and Kirkby 2003). In Spain, 3 per cent of 3.5 million ha of the irrigated area is affected and 15 per cent is at risk (Amezketa 2006).

5.7.3.2 Desertification

The Oxford English Dictionary defines desertification as the 'process by which fertile land becomes desert, typically as a result of drought, deforestation, or inappropriate agriculture'. Drylands, which cover approximately 40 per cent of the land's surface, support 2 billion people, 90 per cent of whom are in developing countries and are particularly vulnerable to desertification. Many dryland regions have experienced a rapid change in vegetation cover and changes in plant community composition, local hydrology, and soil properties, with a consequent loss of ecosystem services that pose serious threats to sustainable livelihoods (D'Odoricho et al. 2013). Desertification results in bare soils (vulnerable to erosion by wind or water); loss of soil functions (water-holding capacity, nutrients); increased salinity; and changes in plant species composition (from perennial to annual species or from palatable to unpalatable grasses). It has been estimated that desertification affects 25 per cent of the world's land surface, which supports 20 per cent of the world's population (D'Odoricho et al. 2013).

Although climate change is likely to play a strong role in desertification through changes in temperature and precipitation, agricultural practice also plays a role including satisfying demand from a growing population; overgrazing and the intensification of previously extensive nomadic livestock systems; deforestation; deep ploughing; and poor water management (such as unsustainable irrigation practices).

Lester Brown (2006) has given examples of desertification from across the world. In Afghanistan, in the Sistan Basin, more than a hundred villages have been buried by windblown dust and sand; in China, deserts are expanding by 3,626 km each year and 400 million Chinese people are affected by seasonal dust storms; and in the Yemen, 97 per cent of land exhibits some degree of desertification.

5.8 Tools for sustainable soil management in agroecosystems

There are a number of soil management techniques that can mitigate soil degradation or conserve soils that are in good condition. These include a variety of integrated agriculture methods that have multiple benefits including to soil, such as reduced or no tillage, increasing crop rotations, agroforestry, and intercropping. These are outlined in Chapter 8. Some examples of very specific measures for soil management are outlined in more detail below. A relatively innovative method of soil management is the use of biochar (see Box 5.5).

5.8.1 Conservation tillage

Conventional deep ploughing can lead to soil degradation through the destruction of soil structure and low levels of SOM. Conservation tillage is practised with the aim of reducing this effect and the term covers any tillage practice that reduces disturbance to soil surface, helps organic matter accumulate, and reduces the negative impact on biodiversity. Conservation tillage includes direct drilling (no tillage) and minimum tillage (i.e. fewer cultivation passes at a shallower depth

Box 5.5 Innovation: biochar as soil ammelioraton

Eroded soils can be mitigated by adding SOM, such as compost. However, the utility of compost is limited in terms of C sequestration and long-term C conservation. A more C-efficient method to increase SOM and simultaneously conserve C is incorporating biochar. Although the original use of biochar is attributed to pre-Columbian Amazonians, it is only recently being adopted as soil amelioration in modern agriculture and is considered a novel technique. Biochar is charcoal and is produced by heating organic matter in anaerobic conditions in a process known as pyrolysis. Biochar is composed of recalcitrant C and degradable C and includes ash as a by-product. The principal difference between biochar and other organic matter, such as compost or manure, is the large proportion of aromatic C and specifically the presence of fused aromatic structures which are responsible for the high stability of biochars (although the mechanisms are not yet fully determined) (Lehmann et al. 2011). The recalcitrance of biochar is due to its relative resistance to microbial decay, although the rate will partly be determined by the nature of the biotic community and the metabolic processes that influence the speed of decay (Lehmann et al. 2011), as well as the source of feedstock, the production processes, and the environment to which it is added (soil type, climate, etc.).

The potential of biochar to impact soil biota is an emerging area of research identified as a priority (Tammeorg et al. 2016). Current evidence is thin and characteristically the results are inconsistent, with positive and negative effects depending on soil type, study organism, and feedstock of biochar. With this caveat in place, broad trends can be identified. The bulk of the published evidence for soil biota focuses on earthworms and microbial communities. Earthworms have been shown to favour soils amended with biochar. There is evidence that biochar acts as a substrate for microbial activity, particularly in soils with little organic matter (Gomez et al. 2014), and can positively stimulate production of macrofauna such as Collembola and Enchytraeids (Marks et al. 2014). However, it is important to note that this effect is strongly dependent on feedstock, as some feedstocks inhibit growth (Marks et al. 2014). In comparison with compost, biochar is of less value to microbes and breaks down more slowly to release nutrients. It has been suggested that using compost and biochar in tandem can maximize the short- and long-term benefits subsequently, although soil fauna may react to potentially toxic elements of biochar (Lehmann et al. 2011). The impact of biochar on crop yields is mixed, with evidence collected from multiple studies suggesting that biochar does not improve yields in fertile soils such as in temperate regions, and only seems to work for yields in poor tropical soils (Jeffery et al. 2017). The sustainability of biochar is an important consideration, resting largely on the sustainability of different feedstocks.

than conventional ploughing) (Holland 2004). It is practised mostly in arid and semi-arid areas, in the USA and Latin America in particular, although it is also employed in Australia and South Africa and has been increasingly adopted in Europe. It has been demonstrated that conservation tillage can reduce erosion and increase SOM, which in turn increases carbon sequestration and develops a richer soil biota which improves nutrient cycling, while crop residues provide resources for insects, birds, and small mammals. Soil type can have an impact on the success of conservation tillage strategies; research has shown that in clay soils it can reduce the tendency of heavy clay soils to form dry cloddy seedbeds that makes sowing difficult but that in unstructured light soils there is a danger of compaction (Morris et al. 2010). A recent study in France that examined long-term effects showed that conservation tillage did not confer the benefits hoped for and that although the upper surface benefited from improved structure and root mass, a compacted layer deeper in the soil formed, preventing longer roots from penetrating (Peigné et al. 2018). Weed build-up can also be an issue under reduced tillage, particularly in no-tillage systems. This has been shown to be a potential issue in a variety of systems across Europe (Soane et al. 2012), and elsewhere including in rainfed rice systems (Chauhan and Johnson. 2009), meaning that the potential soil structure benefits could be offset by the impacts of other weed control, such as increased herbicide applications. Careful management, taking into account timing of cultivation and soil type, is important and more research into the long-term effects of conservation tillage remains necessary.

5.8.2 Green manures and cover crops

Green manures and cover crops are fertility-building crops that are grown for the benefit of the soil. Winter green manures are a short-term intervention, usually sown in the autumn and incorporated the following spring. Summer green manures are usually annual legumes or a brassica and sown to provide a burst of fertility. Long-term manures and cover crops can be laid down as leys and may be left in place for 2–3 years. Green manures anchor soil and the plant cover protects soil against erosion. As the manures break down or are incorporated they add organic matter, thereby improving structural and biological properties of the soil, encouraging microbial activity and supporting biodiversity.

As fertility enhancement is a goal, leguminous crops are frequently included for their nitrogen-fixing ability. Legumes fix nitrogen through a symbiotic relationship with rhizobial bacteria which colonize the leguminous roots, forming nodules within which they fix atmospheric nitrogen. In organic systems this represents a major source of nitrogen and in conventional systems this may be seen as a way to minimize fertilizer inputs. However, in order for legumes to successfully fix nitrogen the appropriate rhizobia must be present (different legumes use different strains of rhizobia) and this is not always the case. In this scenario it will be necessary to inoculate the soil. In the nitrogen-rich conditions (i.e. when nitrogen-rich fertilizer has been applied), legumes will not fix much nitrogen. The timing of incorporation is crucial, as there is a window within which the nitrogen is mineralized and available to plants before it is leached away.

Moreover, green manures and cover crops perform other valuable services including stabilizing aggregates and increasing pore size through an extensive network of fine roots; breaking up compacted soil through the use of species with deep tap roots; providing food (via root exudates) for microorganisms which aggregate soil particles; and increasing SOM to improve water-holding capacity (Rayns and Rosenfeld 2010).

5.8.3 Terracing

Terracing is another traditional practice that conserves soil and also water. Employed in hilly areas, terraces are a series of steps cut onto the hillside to enable farming on a flat plane and are often used for crops that require irrigation, such as rice. Terracing prevents soil erosion and run-off and protects shallow soils. Terracing is an ancient agricultural technique which has been used all over the world: in the South American Andes, farmers have used terraces for hundreds of years to grow potatoes; they are used widely over all of Asia to grow cereals and rice; and in the Mediterranean basin, terraces support vineyards, olive groves, and cork oak. Terraces are acknowledged to be an important cultural phenomenon, and are in some cases protected—for example the rice terraces of the Philippine Cordilleras are designated by UNESCO in the cultural landscape category.

5.8.4 Conservation grazing

Conservation grazing is livestock grazing that is managed with the aim of meeting nature conservation objectives, and usually means fewer animals and a less-intensive grazing regime. Grazing management is an important tool in managing biodiversity, as changing stocking rate, species, or breed of animal can drive shifts in plant community and influence vegetation structure, which in turn benefits animals at higher trophic levels. Furthermore, grazing can be tailored to benefit different habitats such as grasslands, heathlands, or woodlands (Rook and Tallowin 2003). In seriously degraded areas, removing grazing completely can be an important restoration technique. Work on the loess soils in China has shown that removal of grazing significantly influenced soil bulk density moisture content and pH. Long-term fencing also led to increases in SOC, total nitrogen in the soil, carbon:phosphorus and nitrogen:phosphorus ratios, as well as soil C and N storage within the top 100 cm of the soil profile (Deng et al. 2014).

6
Water (Hydrology)

6.1 Introduction

This chapter outlines the importance of water for agroecosystems, and some of the biological impacts of water use for agriculture. Water is essential to food production, and the majority of global agricultural systems are rainfed, irrigated, or a combination of the two, with food usually produced using moisture obtained from soils (Rosegrant et al. 2009). Globally, more cropped land is under rainfed systems than irrigated systems, although this varies between crops, regions, and season, and irrigation still forms an essential part of modern food production. Around 70 per cent of fresh water consumed globally is used for agriculture (Koehler 2008). Water scarcity is growing, and this combined with land scarcity is projected to increasingly constrain food production growth (Rosegrant 2009).

6.2 Importance of water in agricultural systems

Water is one of the most important requirements of agriculture. It is the main component in living cells, and in plants water provides the structural support (or turgidity) to cells, and hence to the whole plant; it transports essential nutrients, is used in photosynthesis, and can help regulate leaf temperature. Most of the nutrients that plants need are taken from the soil or growing medium and transported in water.

In animals, like plants, water helps regulate temperature, but also aids growth, reproduction, lactation, digestion, metabolism, and hydrolysis of fats and carbohydrates. Water helps lubricate joints, cushions the nervous system, and can aid eyesight (Schlink et al. 2010).

Agricultural systems vary in the amount of water that they need. Factors influencing this may include climatic variations; higher temperatures, which increase transpiration in plants for example; and other abiotic factors such as soil type and underlying geology.

Plant species and varieties vary in their water requirements and in their responses to water availability. For example, plants with fewer leaves and stomata have slower

The Biology of Agroecosystems. Nicola P. Randall and Barbara Smith, Oxford University Press (2020).
© Nicola P. Randall and Barbara Smith 2020. DOI: 10.1093/oso/9780198737520.001.0001

rates of transpiration, and plants with deeper root structures can access water that those with shallower root systems would not be able to. Even within species or varieties, the water requirements of individual plants vary under different conditions and during growth; transpiration rates of plants change under higher temperatures and lower humidity, and the amount of water needed varies at different lifestages. Some may tolerate drought as long as they have plenty of water early in their growing season. The final water content of different crops also varies, with vegetables generally having a higher final water content than cereals for example.

Similarly, livestock water requirements are impacted by the type of livestock, climatic conditions, the age and stage of production, and other biological factors such as whether or not they are lactating. Dairy animals, producing milk, need more water than their beef counterparts, for example. The water content of an animal's feed also influences how much additional water it will drink.

The amount of water required to sustain a population can be described as its 'water footprint', and the idea can be extended to the amount of water needed to produce commodities or services, termed 'virtual water' (Hoekstra and Chapagain 2007). Schlink et al. (2010) adapted virtual water figures collated by Hoekstra and Chapagain (2007) for a variety of agricultural commodities. They used these to calculate how many litres of water would be required to produce 1 kg of protein. The findings were highly variable between different products, with soybean requiring 8,952 l/kg compared with 24,464 l/kg for wheat and 38,241 l/kg for rice. Livestock products generally have greater water requirements than plant products, with milk and chicken requiring similar amounts of water per kilogram of protein at 40,896 l/kg and 40,174 l/kg, respectively. By far the most water-intensive product recorded was beef, which on average required 175,631 l/kg. These global figures hide broad variations, however. When different countries were compared, eggs, for example, required 73,067 l/kg in India, but only 13,476 l/kg in Italy, and goat meat varied from 25,522 l/kg in Japan to 102,208 l/kg in Mexico. It is not only climate that contributes to these variations in water consumption per kilogram of protein, but also the technology that is used and the relative yield from different systems. (Hoekstra and Chapagain 2007). Animal products from grazing systems generally have a lower water footprint than industrial systems, although these are usually still less water efficient (with relation to obtaining protein, calories, and fat) than crop systems (Mekonnen and Hoekstra 2012).

6.3 Water use in different agricultural systems

The type of agriculture practised in any region is strongly influenced by climate, with water as a major limiting factor. For example, nomadic pastoralism has long been a feature of semi-arid regions, particularly in parts of north and west Africa and central Asia. In these low-rainfall regions, arable land was historically less viable, and pastoralists moved their livestock around in order to find food and water, often seasonally. Many pastoral systems are now declining due to factors such as land-use and political changes. Agricultural expansion is also a factor in some regions, and often it is changes to water management such as the introduction of irrigation that can facilitate this (see Behnke and Kerven 2013, for example).

6.3.1 Rainfed systems and irrigated systems

Rainfed agricultural systems are those that rely on precipitation for water. They do not have permanent irrigation systems (although there may be irrigation using stored or redistributed rain water). Low or erratic rainfall can restrict agricultural practices. In hotter regions, rainfall may not be adequate for successful crop growth, and so the most productive areas of rainfed agriculture tend to fall in temperate regions of Europe and North America, followed by the subtropics and humid tropics (FAO 2011).

Irrigated systems are those where water is applied to crops using water from man-made means such as pipes or canals. This enables farmers to provide water when rainfall is scarce or erratic.

Portmann et al. (2010) used monthly datasets to investigate the amount of crop harvested from irrigated and rainfed systems for twenty-six different crop classes in the year 2000. Harvested crops included in the study totalled 13 million km^2/year. Of this area, only 18 per cent of the total cultivated land was irrigated, although the cropping intensity was greater on the irrigated area, with the share of the irrigated harvested area at 24 per cent of the total. The percentage of crop irrigated varied greatly by crop type, with rice (62 per cent), cotton, and sugarcane (both 49 per cent) most likely to be irrigated, with others such as cocoa and coffee almost completely rainfed. Asia had the highest amount of harvested area that was irrigated at 41 per cent, compared with only 7 per cent for Europe, although the authors reported large within-region variation.

Irrigation systems can receive their water from precipitation, but there is increasing focus on ways to improve or expand irrigation measures, due to increased yields from irrigated crops, which may be as much as 60 per cent greater, and irrigated agriculture is expected to provide over 50 per cent of the growth in cereal production over the period 2000–2050 (Rosegrant et al. 2009). Permanent irrigation systems usually receive their water from surface or ground water. In 2010, 62 per cent of irrigated area received its water from surface waters, with the rest coming from ground water, although this is highly variable; at that time, 88 per cent of the total irrigated area in the Arabian peninsula came from ground water (FAO 2011).

Research summarized by Rosegrant (2009) indicated that crop water consumption is expected to increase at 0.7 per cent annually between 2000 and 2050. He also summarized that irrigated harvested area is expected to increase by 0.24 per cent annually during the period 2000–2050 (although some of this irrigated area is likely to get water from precipitation rather than withdrawal as there is an expansion of irrigations into wetter regions) and rainfed harvested area may increase by

0.13 per cent per year. In Sub-Saharan Africa irrigated agriculture is projected to double.

6.3.2 Water-based agricultural systems: hydroponics, aquaponics, and aquaculture

Hydroponics, aquaponics, and aquaculture are all production systems that have water as a basis rather than a factor in their design.

Hydroponics is a method of growing plants without soil as a growing medium. Instead plants are fed on a water solution containing mineral nutrients, either by being placed in an aeriated solution or via continuous application of a mist of solution. There may sometimes be other growing media, such as fibres or mosses, but there is contention as to whether these systems should be called hydroponics or merely soil-less systems. Although often thought of as a relatively modern concept, it is thought that hydroponic-type systems were actually used in the hanging gardens of Babylon and in the floating gardens of the Aztecs of Mexico (Jones 2016). Hydroponics became widely used commercially in the 1980s, mainly in greenhouses, and by 1995 over 24,000 ha of greenhouse vegetables were grown using hydroponic systems (Jones 2016). Use of hydroponics enables plants to be grown at high density but with a reduced chance of disease from soil organisms, but the plants need artificial support. This highly artificial system means that most elements of the production are more highly controlled than in natural environments. Growing conditions can be monitored and adapted to improve productivity.

Aquaculture is the farming of aquatic organisms either in freshwater environments or in the sea (mariculture). Fish, molluscs, aquatic plants, and algae farming are all considered forms of aquaculture. As with terrestrial agriculture, in aquaculture the organisms and their environment are managed for production, which makes it distinct from fishing, which is a form of natural harvesting. As in most agroecosystems, aquaculture systems and their level of intensity are highly variable. For example, fish may be caged or penned, or allowed to swim free in ponds, and shellfish may be cultured on rafts, vertically, or on sea beds. Both fish and crustaceans are regularly farmed in rice paddies in China (Cao et al. 2007). Artificial feeding may or may not take place. China is by far the largest aquaculture producer in the world. Its history of aquaculture dates back more than 2,000 years but has rapidly developed since the 1990s (Cao et al. 2007).

Aquaponics is a combination of aquaculture with hydroponics or other forms of agricultural plant growth in an integrated system. For example, in a freshwater system that contains both plants and fish, the fish waste provides an organic food source for the plants and the plants naturally filter the water for the fish. Floating wetlands are often incorporated into aquaponic systems. These floating wetlands rise and fall with the water level as it changes, and this can offer benefits over sediment-rooted wetlands, because plants die off less frequently. This in turn releases fewer nutrients into the water. Figure 6.1 shows an example aquaponics system.

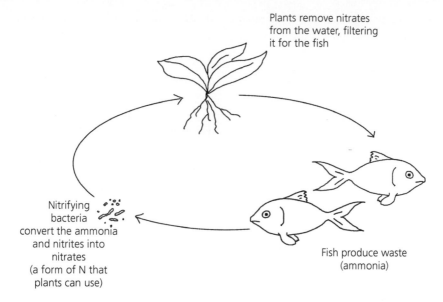

Plants remove nitrates
from the water, filtering
it for the fish

Nitrifying
bacteria
convert the ammonia
and nitrites into
nitrates
(a form of N that
plants can use)

Fish produce waste
(ammonia)

Figure 6.1 A simplified aquaponics system, where the presence of fish, plants, and microbes allows an integrated system, whereby the fish waste provides an organic food source for plants, with support from the nitrifying bacteria, and the plants provide a natural nitrogen filtration system for the fish. Both the fish and the plant products can be used as agricultural outputs. Original line drawing: Nicola Randall

6.4 Issues and challenges for water in agroecosystems

Management of agricultural systems can have biological issues that are more far reaching than within the farmland system itself. This is particularly the case with issues relating to water ecosystems. Water for agriculture is often imported into agricultural systems from natural environments, impacting on the wider water cycle and the environments that water has been removed from. Water exported from agricultural systems often contains pollutants from the agricultural systems, which can move rapidly in mobile aquatic environments. Some of the biological issues relating to water use in agriculture are illustrated in Figure 6.2.

6.4.1 Availability of water in agricultural systems

For the growth of healthy crops, most water-related issues are caused by either too much soil water, causing waterlogging, or too little.

Plant roots take up moisture via hairs on their roots, and release it through their leaves via transpiration (Figure 6.3). In soil, the amount of water that is available to crops and other plants (plant-available water) is influenced by the soil pores (which in turn are influenced by soil structure, biology, and management—see Chapter 5) and the plant roots themselves. Finer soils tend to hold more water as

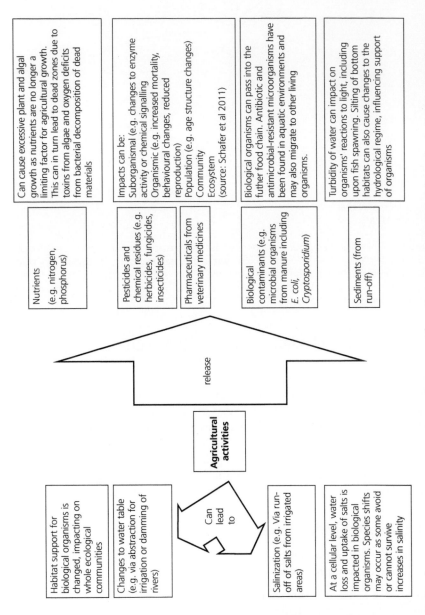

Figure 6.2 Biological issues in aquatic ecosystems caused by agricultural activities (from information in Holden et al. 2015).

they have smaller pore sizes and larger surface areas than particles in coarser sandy soils, and so water is less easily drained away by gravity. Very fine soils (such as clay-based soils) have high quantities of water, but this is of limited availability to plants as water is bound too tightly to the soil in the small pores. Similarly, if soils become too compacted and the pores too small, water cannot infiltrate into the layers at all. In contrast, coarse, sandy soils may not hold enough water, meaning that the top layers can dry out very rapidly.

The key soil pore sizes that influence water availability to plants are macropores (those that are > 75 μm), mesopores (30–75 μm), and micropores (5–30 μm) (Soil Science Society of America 2008). Of these, the most important are mesopores, which can store water that is useful to plants, and micropores, in which water is usually immobile but may be able to pass solutes to plants via diffusion. Macropores are too large to hold water so are usually air filled when soil is at field capacity (a term used to describe soil state when the water content has reached a nearly constant value following wetting by rain or irrigation and downward drainage has nearly ceased in a free-draining soil (Kirkham 2005)). See Figure 6.3.

The life within and around soil is also essential in water storage regulation. Plant cover, soil organic matter, and the biotic community all have an influence. For example, plants and surface litter trap sediments, and invertebrates that move between soil and litter layers influence water movement within and over soils (Power 2010). Soils with more organic matter are also able to hold more water; as well as influencing the structure of the soil, the particles of organic matter have a charged surface that can attract water. An increase in organic matter from 0.5 to 3 per cent has been shown to increase the available water capacity of various soil textures (Hudson 1994).

6.4.1.1 Aridity, drought, and agricultural production

Lack of water impacts on crop yields, and in 2006 around 28 per cent of the globe was classified as arid under the Köppen–Geiger climate classification (Kottek et al. 2006). Water can be variable in all climates at some point, however. Drought is generally thought of as a period of months (or even years) in which precipitation is less than the annual average. For agriculture, this can be considered a period where a lack of surface water resources leads to declining soil moisture and consequent crop failure (Opiyo et al. 2015). Drought-induced yield loss is considered to be one of the greatest impacts on agricultural yields, with global wheat crops on average experiencing a 20 per cent yield reduction with a 40 per cent drop in water and maize experiencing nearly 40 per cent yield losses (Daryanto et al. 2016). Water stress in plants causes wilting, closure of stomata, decrease in cell growth, and, if severe, an arrest of photosynthesis and death (de Oliveira et al. 2013. Crops, as with other plant species, have different sensitivities to drought at different growing stages, with small seedlings often particularly vulnerable as both photosynthesis and growth are reduced. It is not necessarily plant survival that is impacted, but other factors related to growth and development that are inhibited. In wheat,

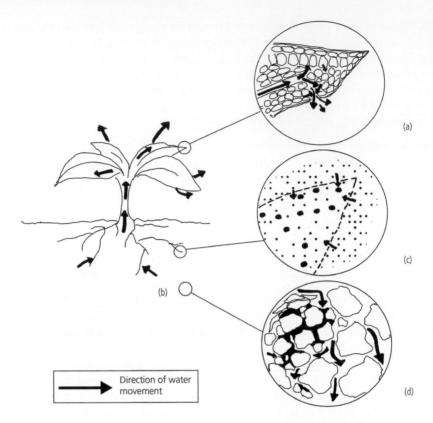

Figure 6.3 The process of transpiration. (a) Water in the plant mesophyll cells moves into the intercellular spaces, and then evaporates through small gaps in the leaf surface (stomata) via a process of diffusion. (b) This transpirational pull causes water to move from the roots to replace water lost from the stems and leaves in transport tissue known as xylem. (c) Osmosis takes place in plant root hairs. Water particles pass through a semi-permeable membrane from the soil to the plant when the water concentration is higher and the salt concentration lower in the soil than in the plant. This root pressure contributes to upward movement of water through the xylem. (d) Differences in soil particle size and shape influence the movement of water in soil and so impact on the availability of water to plants. If the spaces between soil particles are too large, the water will drain away. If particles are too compressed, water may not be able to penetrate into the gaps between them. Original line drawing: Nicola Randall

for example, drought can cause yield reductions of up to 92 per cent. The amount and quality of grain produced by each plant can be inhibited by drought during the reproductive development stages, and may range from sterility or malfunctions caused by drought during meiosis or microspore stages causing reductions in the number of flowers that initially develop on the spike, or premature abortion of florets if drought occurs slightly later in development (Senapati et al. 2018).

In livestock or pastoral systems, dehydration and starvation of animals are serious impacts of drought, although increased disease can also be a factor in livestock mortality under drought conditions. One study in Ethiopia, for example, attributed

60–100 per cent of excess livestock mortality to starvation/dehydration and up to 30 per cent to disease following drought conditions (Catley et al. 2014).

6.4.1.2 Salinization and agricultural production

Salinization is the process of the accumulation of water-soluble salts in surface soils. This takes place when the dissolved minerals that are present naturally in water are left behind when water evaporates. Salinization takes place in coastal areas when land upstream is drained and salts are drawn into inland soils, or can be caused by irrigation on land that is naturally dry, such as on desert margins, where irrigation causes mobilization of salts in the soil. In plants, soil salinity can inhibit plant growth for two reasons: (1) accumulated salts in the soil solution can reduce the ability of plants to take up water, as it interferes with the osmotic process, which causes water stress (see section 6.4.1.1); and (2) salinity can cause toxicity to plants, where salts are actually taken up by plants and can cause nutritional imbalances and specific ion effects, which can injure cells and cause leaf fall (de Oliveira et al. 2013). In small amounts, salinization can impact on yields by changing the metabolism and productivity of plants; in larger amounts, the soil can become toxic and salinization may even make it unusable. Also see Chapter 5, section 5.7.3.1.

Salinization can also impact on livestock production due to reduced quality or amount of forage or suitable grazing areas. Some livestock are able to graze on halophytic (salt-tolerant) plants, but these tend to be low in both nutritive value and edible biomass. High-salt diets are thought to impact on all stages of animal production, from conception to gestation, through weaning, early growth, and finishing (adding weight and optimizing fat and muscle pre-slaughter) (Masters et al. 2006).

Specific impacts of salts on livestock include depressed appetites, water retention, increased pulse and respiration rates, and changes to the digestive function. Sodium, potassium, and chloride are the salts that cause most issues in animals, but tolerance varies, with sheep and cattle able to tolerate a higher percentage of sodium chloride in the diet than chickens. Salinization effects also vary depending on whether the high salt intake comes solely from feed or whether it comes from both food and drinking water. If supplied with unlimited fresh water, many animals can cope with high salt content in feed by increasing their water intake, which allows the kidneys to increase their salt-excreting capacity. Where drinking water is limited or where both water and feed are contaminated with salts, this change to kidney function is not possible (Masters et al. 2007).

6.4.1.3 Waterlogging and agricultural production

It is not only too little water that can impact on plant growth; too much water can cause issues too. Waterlogging occurs when soil is saturated with water, which results in anaerobic conditions in the plant root zone, resulting in a lack of oxygen necessary for plant growth and survival. The majority of crops are negatively affected by waterlogging For example, it has been demonstrated that oxygen-poor conditions limit nutrient uptake by barley plants (Steffens et al. 2005) and in wheat

Figure 6.4 Waterlogged rice paddy. Photo: Barbara Smith

it has been shown that even short-term (3–7 days) waterlogging can have a nega-tive effect on root development in young plants (Malik et al. 2002). Waterlogging can also result in reduced water uptake as roots require oxygen for normal cell function, and prolonged waterlogging can lead to plant death. Furthermore, in irrigated conditions, waterlogging can lead to salinization as the waterlogged soils prevent the imported salt from leaching away.

Waterlogging can occur as a natural process due to excessive rain or flooding. In agroecosystems poor irrigation management or soil compaction due to pressure from vehicles, shallow tillage, or trampling by animals can also cause waterlogging.

Not all crops require aerobic respiration. Rice, for example, is tolerant of water-logged conditions, as it has a process for dealing with oxygen depletion in the rhizosphere. Waterlogging induces cell death in the roots, resulting in gas spaces in soft plant tissue (known as aerenchyma), allowing oxygen to be transported from the leaves down to the roots (Yamauchi et al. 2017). Rice fields are routinely flooded not because it is necessary for the rice to grow, but because it is used as a method of weed control, as most other plants don't have the capacity to survive the inundated conditions (Figure 6.4).

6.4.2 Manipulation of natural water systems for agriculture

In order to enhance agricultural production, humans have long manipulated natural systems. As water is one of the key limiting factors for food production, manipula-tion of water for agricultural production is an issue for the natural biology of sys-tems. This manipulation usually involves moving water from one place to another.

6.4.2.1 Water abstraction for agriculture

The extent of agriculture has had a significant impact on freshwater resources, affecting surface run-off through large-scale deforestation. Agriculture also abstracts large amounts of water, largely for irrigation, and global water withdrawals now account for approximately 10 per cent of the global renewable resource (Foley et al. 2005). As a result, many large rivers have greatly reduced flows (particularly in semi-arid areas). Some rivers, (particularly in India and China) no longer reach the sea, which has resulted in coastal salinization and changes to coastal ecosystems. Inland lakes can also be impacted; for example, between 1979 and 2001 the volume of Lake Chapala in Mexico depleted by 80 per cent (FAO 2011). Groundwater extraction has led to the depletion of aquifers and to declining water tables in many regions (Foley et al. 2005). Box 6.1 shows how reliance upon water abstraction for agriculture varies globally.

Box 6.1 Consumption of water for agriculture—green, blue, and grey water

Water availability is a limiting factor for agriculture, particularly in arid areas. The amount of global water stays constant, but the availability and usability of water can vary. Water that is withdrawn (or abstracted) from surface waters, such as rivers and lakes or ground water, and used for irrigation is sometimes known as blue water. Although around 70 per cent of the water extracted from surface and ground water is used for agriculture globally, only about 9 per cent of internal renewable water resources are withdrawn in this way, and this figure varies greatly. In Europe, water is generally readily available, and most systems are able to use water from precipitation (or green water), most of which is stored as soil moisture, so Europe withdraws only 6 per cent of blue water, with less than 30 per cent of that being used for agriculture. In contrast, parts of the Middle East, northern Africa, and central Asia have very little available water that is not already withdrawn, and 80–90 per cent of the water that is taken is used to support agriculture (FAO 2011). Table 6.1 shows the percentage of withdrawn water that is used globally. In some water availability considerations, polluted or 'grey' water is also included. Grey water is defined as the volume of fresh water required in order to assimilate the load of pollutants, given natural background concentrations and existing water quality standards (Mekonnen and Hoekstra 2010).

Table 6.1 Percentage of withdrawn water used for agriculture in 2010

Africa		Americas		Asia		Europe		Oceania	
Africa	81	Americas	48	Asia	81	Europe	25	Oceania	65
Northern	84	Northern	40	Middle East	84	Western and Central	27	Australia and New Zealand	65
Sub-Saharan	79	Central	59	Central	89	Eastern and Russian Federation	21	Pacific Islands	59
		Caribbean Southern	71	Southern and Eastern	80				
World			69						

6.4.2.2 *Drainage of land for agriculture*

Drainage of land to improve its suitability for agriculture is another way in which natural water processes are disrupted. Historically, some low-intensity agricultural systems in wetland areas have taken place with little or no drainage or additions of pesticides and fertilizers, resulting in ecologically diverse systems. The moist grasslands and herbaceous fens of Europe were used for grazing and haymaking but stayed flora and fauna rich, traditional rice cultivation allowed fish, macro-invertebrates, and waterfowl to thrive, and floodplains were grazed but not engineered to reduce flood events (Verhoeven and Setter 2009).

In other regions, drainage of land for agriculture has been taking place for thousands of years. In Mexico, for example, the Aztecs are thought to have created series of canals (or chinampas) and raised beds in shallow lakes in order to grow food as far back as the 1300s. With the exception of water-loving crops such as rice, poorly drained soils are usually unsuitable for intensive agriculture (see section 6.4.1.3). Waterlogged or frequently flooded soils often have high organic matter content, and this can make them more desirable areas for agricultural expansion than other marginal regions such as uplands that often have poor thin soils and may be difficult to access. Reclamation of wetlands for agriculture usually involves a combination of drainage and soil improvement (Verhoeven and Setter 2010), which changes the character of an ecosystem and all the natural processes and functions that take place within it.

Since the early nineteenth century, intensification has taken place alongside engineering to regulate or drain wetland areas. It is estimated that in North America, Europe, and Australia, over 50 per cent of wetland areas have been lost, largely due to agriculture (Millennium Ecosystem Assessment 2005). In South East Asia, deforestation of approximately 45 per cent of the region's tropical peatlands occurred over just 50 years in the latter part of the twentieth century, most of which were also drained and are now in agricultural use. Increasingly the crops being grown need more fertilizer use and deeper drainage. This contributes to biodiversity declines and to carbon emissions.

6.4.2.3 *Hard landscaping for agriculture and impacts on water distribution*

Hard landscaping in agriculture is an increasing phenomenon. This can range from permanent indoor housing for livestock to polytunnels and glasshouses for agriculture and horticulture. Almeria in southern Spain is infamous for its 135 square miles of glasshouses (see Plate 5). Although much research has been carried out on water use within these agriculture systems, there is little research on how increases in permanent or semi-permanent structures designed for agricultural production influence the wider water cycle. However, the influence of impervious surfaces on run-off and water balance has been well studied in urban areas and is known to impact hydrogeological cycles, for example by changing the rate of water discharge to watercourses (Shuster et al. 2005).

Reflectivity (such as from glasshouses) can also impact on evapotranspiration, and this is something that may become an increasing issue if the trend towards increasing use of glasshouse and indoor housing for livestock continues.

6.4.3 Water pollution from agriculture and the impact on fresh and coastal waters

Water quality has been degraded by intensive agricultural practices through the leaching of agrochemicals and nutrients into ground water and waterways along with sediment from eroded soils. Agriculture is responsible for the majority of the nitrogen and phosphorus loads in freshwater systems (Bennett et al. 2001).

Nutrient applications in excess of plant needs, together with increased run-off, can contribute to water quality decline. It is not just the application of nutrients and pesticides to the surface that impacts on these processes but also other management of the land. Compacted soils and reductions in organic matter content lead to increased risks of soil erosion and run-off. This phenomenon has been much studied in Europe, where it is estimated that each year up to 200 million m^3 of sediment is dredged from European rivers (SedNet 2004). Studies in the UK have estimated that agriculture is the source of 28 per cent of phosphates, 70 per cent of nitrates, and 76 per cent of sediments in rivers, and UK agricultural catchments have elevated levels of bacterial pathogen counts, which are also likely to be from agricultural systems (Collins et al. 2015; Edwards and Withers 2008; Kay et al. 2008). Pesticides can also leak into watercourses. One study found 500 chemicals in four European river basins (the Elbe, Danube, Schelde, and Lobregat), forty of which were at levels harmful to organisms and 75 per cent of which were pesticides (Von der Ohe et al. 2011). Nitrogen is strongly water soluble and any excess can be easily washed into watercourses. Galloway et al. (2004) estimated that about 20 per cent of nitrogen-based fertilizer applied to agricultural systems moves into aquatic ecosystems.

Nutrients can cause eutrophication (enrichment) in waterbodies. This in turn leads to faster growth of some plants and algae. This has a number of impacts: diversity is decreased as the most competitive plants thrive at the expense of others; the increased biomass and bacterial activity as the plants die off can lead to oxygen depletion (hypoxia) which in turn can lead to fish deaths; and the algal blooms shade out the water and can themselves be toxic to wildlife. This can impact both fresh waters and coastal marine ecosystems (Power 2010). For example, in marine systems, high agricultural run-off has been associated with large phytoplankton blooms in vulnerable areas of the ocean (Beman et al. 2005). In coral reefs, where nutrient enrichment has been shown to impact the tolerance of corals to heat and light stress and their susceptibility to bleaching, an increase in phosphates has also been shown to reduce skeletal density so that corals become more brittle (D'Angelo and Wiedenmann 2014).

6.4.3.1 Pollution from aquaculture

As aquaculture takes place directly within water environments, the impacts on water quality can be more immediate than those from land-based agricultural systems. Aquaculture can impact on wider ecosystems in a variety of ways: eutrophication through the build-up of waste materials, and the potential introduction of alien species or of pathogens into new environments. A wide range of chemicals (e.g. pigments, disinfectants) are also used, which can impact on natural ecosystems, and a particular concern is the use of antimicrobials causing pathogen resistance. Farmed fish often also need to consume wild fish as part of their diets, so harvesting fish for human consumption can have broader ecosystem impacts. Pollution is not the only issue connected with aquaculture systems; other biological impacts of the process are common. For example, over a third of global mangrove forests were lost in a 20-year period covering the late twentieth and early twenty-first century; of this, a third was converted to shrimp ponds. The decline of these mangroves has resulted in losses of a variety of ecosystem services, including habitat provision for wildlife and fish nurseries, flood control and coastal protection, and water quality provision through sediment trapping and natural filtering (Primavera 2006).

6.5 Tools for water management in agroecosystems

6.5.1 Sustainable water use

One of the key ways to reduce the impacts of agriculture on the water environment is to improve the sustainability of water use.

Use of recycled water (e.g. for irrigation) is one way to reduce the reliance on blue water. Approximately 1 per cent of irrigated water came from either recycled or desalinated water in 2010 (FAO 2011). The use of recycled water in irrigated systems is rising particularly around urban areas, where wastewater treatment methods are being developed and drainage water can be re-used. Both of these methods have to be carefully managed, as waste water, although high in nutrients, can contain other contaminants, and drainage water can become salinated as salts in the soil are leached into the drainage water. Desalinated water is less common in agriculture, but is sometimes used in high-value crops or where no other water source is available. The use of desalinated water is increasing, as costs of desalination reduce while costs of groundwater extraction rise.

Another way to reduce the negative impacts of irrigation is to reduce water losses from irrigation by targeting the irrigated water more effectively. One method for this is the use of subsurface drip irrigation systems. Irrigation pipes are laid below the soil surface. Water is targeted to the crop root zone, rather than being lost elsewhere in the soil, and as the pipes are below the surface, evaporation and run-off are minimized. The reduction in evaporation is particularly valuable in hot, arid, or very windy conditions. Subsurface drip irrigation also means that both farmworkers and crops have a reduced risk of coming into contact with any contaminants if waste water is used within the irrigation system.

Rainwater harvesting is another method to reduce the reliance on blue water. Run-off and evaporation can be reduced by effective harvesting of rainwater either in the storage areas or in the soil profile itself. Rainwater can be caught in dams, diverted in simple ditches, or stored in tanks. Terracing of slopes is also a form of rainwater harvesting, and works by stopping surface water run-off (and associated soil) from moving downslope.

6.5.2 New crop varieties

Species and varieties of crop/livestock vary in their water requirements and in their tolerance to extremes in water availability. Some crops are more resistant to saline conditions, which can be useful where saline drainage water is mixed with fresh water, or where soils suffer from low levels of salinization. Examples of crops that can tolerate more saline conditions include durum wheat, triticale, and barley, when compared with rice and maize. The growth rate of wheat is less impacted by salt than maize, which is less impacted than rice. Yields are also impacted in varying amounts, and this often varies between cultivars. Some vegetables crops may even benefit from some salinity; tomatoes and melons, for example, may have increased sugar levels when irrigated with slightly saline water (FAO 2011). These natural variations between species and varieties can be utilized by scientists. Drought-resistant cultivars and flood-tolerant varieties of crops are constantly being developed.

6.5.3 Land and soil management

A variety of land management options can be used to help reduce water losses (and associated pollutants) from agricultural systems, and many of them involve methods to simulate more semi-natural systems. Planting of cover crops (crops planted to stabilize soils and reduce run-off from other farmed areas) or of permanent vegetative strips adjacent to productive land can provide ground cover and root systems, reducing surface and subsurface run-off and stabilizing soils. Excess nutrients are also taken up by the uncropped plants, reducing the likelihood of them reaching watercourses. Vegetative strips have the added benefit of providing shelter and food resources for invertebrates and other farmland wildlife, so increasing biodiversity and even potentially supporting related ecosystem services such as pollination and pest control. Other management tools such as controlled farming traffic (controlling the type of machinery on land to reduce soil compaction) and changes to tillage practices are designed to improve soil quality, but can also reduce run-off and improve infiltration and water quality. One of the key soil management methods to improve water retention is the incorporation of organic matter. Some of these methods are explored further in Chapter 8.

6.5.4 Anti-transpirants (waterproof crops)

One innovation specifically designed to reduce water losses in crops in water-poor or hot regions is the use of anti-transpirants. These waterproofing sprays are applied to the leaves of crops and reduce the quantity of water lost through the leaves of the growing plant through the stomata as they reduce the rate of transpiration. The crucial stage for application is around the time of pollen production. If the pollen grains are still developing and the plant becomes short of water, the pollen grains die. In cereals such as wheat, this means that the grains then can't develop and the yield will be poor. Thus, spraying droughted crops before the pollen grains are produced can help to increase wheat yield.

6.5.5 Sustainable aquaculture

Feed and feces wastes from aquaculture are key sources of nutrient loading, but developments in feeds have led to lower nitrate and phosphate content in both feed and fish feces. Waste can be reduced further through optimized feeding systems, where new technologies facilitate monitoring and reduction of food waste.

Aquaculture systems that move towards aquaponics or other more integrated systems where the wastes of one type of organism (e.g. fish) become the fertilizer or food for another (e.g. seaweed or filter feeders) can offer a more balanced ecosystem and can help reduce eutrophication. Other ways of reducing waste impacts can be to remove waste products and use them as fertilizers in agricultural or forest systems (Cao et al. 2007).

6.5.6 An integrated approach to water management for agriculture

With water management, as with many aspects of agricultural management, an integrated approach is often the most effective way of managing the system. This may be through small-scale or on-farm changes such as diversification: for example, in Kenya, introducing agroforesty into smallholder systems and diversification of herd composition in pastoral systems, combined with training and education have been strategies used to mitigate drought situations (Thorlakson and Neufeldt 2012; Opiyo et al. 2015). Much larger-scale integration may include regional change influenced by politics and funding. Box 6.2 illustrates how a large-scale integrated approach is being attempted in Lake Taihu in China.

Box 6.2 Lake Taihu, China: an integrated approach to reducing water pollution

Lake Taihu is the third largest freshwater lake in China and is estimated to be the main drinking water source for over 10 million people. Since the 1980s, excess nitrates and phosphates from agriculture, aquaculture, industry, sewage plants, and air pollution have caused accelerated eutrophication, leading to algal blooms becoming a continuing issue in the lake.

In the early 2000s, researchers decided to introduce floating wetlands, designing the largest aquaponics system in the world, at over 2 ha in size. Floating beds of plants such as rice are grown and harvested on the water. The plant roots remain in the water and serve as habitat for native fish, which in turn eat algae. Initially thought to be a success, the project has been impacted by ongoing pollution in the lake. Decomposed cyanobacterial scum, caused by eutrophication, led to a drinking water crisis in 2007, affecting 2 million people (Qin et al. 2010).

The authorities realized that a more integrated approach needed to be implemented in order to address the water quality issues: a variety of targets to reduce pollution have been set, manufacturing plants have been forced to close or have been relocated, monitoring and restrictions of effluents are in place, and new sewage treatment plants have been built. Agricultural targets have included reducing biocides and nitrogen use. Chemical fertilizers have been strictly controlled since 2000 (Wang et al. 2009) and policy options to reduce agricultural pollution further include payments or subsidies for farmers to convert areas close to rivers from arable to trees, to convert to mechanical rice transplanting, and to build biogas digesters. Agricultural extension offices are also given funding to use soil sampling to provide site-specific recommendations on nutrient management (Feng et al. 2012).

7

The Biological Impacts of Globalization of Agriculture

7.1 Introduction

Despite the continuing existence of local subsistence agriculture in some countries, globalization has had a big impact on the way that much of our food and other agricultural products are produced. Agricultural products are traded globally and increasingly food consumption and production are spatially separated so that consumer demand in one place affects which crops are grown in another, thereby linking distant socio-economic systems (Furumo and Aide 2017). In many countries, diversity in the diet is attained through imports rather than locally diverse agriculture, and countries with short cropping windows import foods that are out of season locally. There is also increasing demand for exotic vegetables and unusual varieties. Between 1986 and 2009 the amount of food calories traded in the international market more than doubled and the number of links in the trade network increased by over 50 per cent; in 2014, 23 per cent of food for human consumption was traded internationally (D'Odorico et al. 2014).

Globally there is increasing pressure on agriculture as demand for food increases due to population growth. The global population has increased by 1 billion every 12–14 years since the 1960s (D'Odorico et al. 2014) and is projected to increase to 9 billion by 2050 (Godfray et al. 2010). To date, food production has largely tracked population growth, and increases in yields can be attributed to the intensification of farming via plant breeding, irrigation technology, increased use of artificial fertilizers, and improved pest control as well as expansion into previously uncultivated areas. Furthermore, consumer demands are changing with increasing affluence, particularly with respect to higher demand for animal products, which results in increased greenhouse gas (GHG) emissions.

Since the 1960s, expanding population numbers coupled with changes in diets have driven land-use change (Alexander et al. 2015) and this has had an impact on the identity and diversity of crops grown globally and the scale at which they are grown. Croplands and pastures now cover around 40 per cent of the terrestrial surface. Similarly, the clearing of lands for agriculture and the replacement of

The Biology of Agroecosystems. Nicola P. Randall and Barbara Smith, Oxford University Press (2020).
© Nicola P. Randall and Barbara Smith 2020. DOI: 10.1093/oso/9780198737520.001.0001

natural forests with plantations have resulted in the deforestation of between 7 and 11 million ha of forest globally since the eighteenth century. Despite this, in some areas, improved management has at least stabilized forest loss and increased the standing biomass of tree cover, if not restored the original biodiversity or increased the land area covered significantly (Foley et al. 2005). Pressure on land is likely to increase as our ability to increase yields plateaus and this is exacerbated by the competing demand for biofuels.

Changes to land use are not the only biological impacts of the intensification of cropping to supply global markets; there are changes in nutrient flows (which includes the movement and relocation of nutrients), the movement of species (including crop species, pests and diseases, and invasive species), an increased demand for water, and increased use of chemical inputs (fertilizers and pesticides) that are at least partially a result of globalization.

Further attempts to maintain and increase production have resulted in new methodologies, including the genetic modification of crops to resist pests; the development of elite rhizobia to improve yields; new pesticides to control increasingly resistant pests; and precision farming to optimize the use of inputs. These technologies proliferate across the world through globalized industry and communication channels, facilitated by liberal trade economies. At the same time, knowledge exchange and collaborative development of technology have been globalized and this has led to improved capacity to develop mitigating strategies, although some technological solutions, such as genetic modification of crops, can be controversial.

7.2 Shifting diets and commodities

7.2.1 General trends

Foods and agricultural products have long been traded globally (the ancient Egyptians were known to import spices, and the ancient Greeks and Romans are known to have traded with India for example); but as dietary and other commodity requirements have changed, and demand has grown through the development of international markets, the impact on countries that provide the products has also increased. In recent years, growing affluence and urbanization, in addition to increasing meat consumption, have driven a global transition from traditional diets to those higher in refined sugars, refined fats, oils (Tilman and Clark 2014), and non-essential, often high-value crops. Increasingly, landscapes are simplified in order to specialize and supply the global market with staple and cash crops (crops grown to sell for profit).

7.2.2 Globally traded cash crops

Many crops are traded internationally, but some crops have had a particularly high impact. The farmers of tropical regions have long supplied the global market with cash crops, and in many cases this has had an impact on local agroecosystems by changing the way the crop is produced.

For example, coffee and chocolate, are both forest products which have been grown in plantations in order to produce at scale. In both cases, farmers have been encouraged to select sun-tolerant varieties, to clear land, and grow intensively, resulting in large-scale deforestation and loss of related biodiversity, much of which has gone undocumented. In open conditions, coffee crops have been found to suffer from high weed and insect pest burdens, leading to high pesticide use, and without the cover of shade trees, soils are quickly depleted of organic matter and have suffered concomitant soil erosion (Staver et al. 2001; Hartemink 2006). A study in Indonesia compared shade-grown and sun-grown coffee and cocoa farms and found that canopy height, tree, epiphyte, liana (characteristic vines), and bird species diversity, vegetation structural complexity, percentage ground cover by leaf litter, soil calcium, nitrate nitrogen, and organic matter levels in the organic horizons were all greater in shade-grown farms. In contrast, air and soil temperatures, weed diversity, and percentage ground cover by weeds were greater in sun-grown fields (Siebert 2002). In southwestern Ethiopian montane habitats, the move away from coffee growing in traditional agroforestry systems to relatively intensive monocropping of cereals has resulted in a reduction in biodiversity and soil fertility (Kassa et al. 2017). Despite this growing evidence, shade-grown coffee and cocoa represent a relatively small proportion of global production; in 2010 shade-grown coffee represented 24 per cent of the global area planted (Hernandez-Aguilera et al. 2019). It is predicted that future climate variations, including temperature rises and potential changes to precipitation, will decrease yields and quality in the context of increasing pests and disease, and the area of land suitable for coffee growing in big coffee-producing nations such as Brazil, El Salvador, and Nicaragua may fall by up to 80 per cent. Finding sustainable alternatives is increasingly important and shade-grown coffee is likely to be an important part of the solution (Hernandez-Aguilera et al. 2019).

The current high economic returns for monocropped systems lead to land clearance. This is exemplified by the expansion of rubber plantations in South East Asia, where the demand for rubber more than doubled between 2004 and 2019 (Chiarelli et al. 2018), despite concerns that there will be negative environmental outcomes. Rubber replaces native forest, thereby reducing biodiversity and affecting carbon and hydrological cycles (Qui 2009). It also competes with food production for both land and water (Chiarelli et al. 2018). Qui (2009) reported that in Xishuangbanna (a tropical region of China) the characteristic dense winter fog that provides water to a wide range of rainforest plant species in the non-rainy season has been reduced by the planting of rubber trees. Rubber trees—known as water pumps by the local communities—suck up more water and cause surface run-off, resulting in soil moisture depletion and reduced evapotranspiration. In Xishuangbanna this has led to reduced stream flow and village wells drying up. To put this into context, the crop water requirement (CWR), or the depth of water (in millimetres) needed to meet the water consumed through evapotranspiration, is considerably higher for rubber (1,200–1,400 mm/year) than either wheat or maize (typically 300–600 mm/year) (Chiarelli et al. 2018).

Rubber competes for land with food production at two scales. Much of the rubber production in South East Asia is planted on smallholder farms, displacing local food production. More recently, large land acquisitions are being made by private enterprises, which put further pressure on local food security and intensify production, thereby increasing the impact on the environment (Chiarelli et al. 2018). There have long been calls for more diversified production systems to be reintroduced (Fox et al. 2014), to address both environmental and food security concerns.

Similarly, oil palm plantations have received much negative attention as they expand to satisfy the global oilseed and biofuel markets; palm oil is ubiquitous in many processed foods, detergents, and cosmetics and accounts for 35 per cent of the oilseed market (OECD 2017). The majority of the increase in palm oil production has been in land expansion rather than yield improvement, and Asia has seen the replacement of forest with palm oil monocultures, with implications for biodiversity, ecosystem function, and carbon emissions (Furumo and Aide 2017). Although the expansion of oil palm has led to the loss of primary native forest and its associated biodiversity (Koh and Wilcove 2008), the impacts are not the same in all regions. For example, in Latin America, the trajectory is different and plantations are more likely to replace cattle ranches or banana plantations, with only 21 per cent of the area under palm oil being cleared forest (Furumo and Aide 2017).

Demand and market forces determine which crops are selected for intensive agricultural production. Aizen and Harder (2009) demonstrated that in recent years there has been agricultural expansion of pollinator-dependent crops (including oilseed, nut, and fruit crops), all of which have fed global markets. This has been driven by a higher than average market value for these crops, and a lower intrinsic yield growth for them, which means that to increase yield, larger areas must be planted, placing a burden on global pollinator capacity (Aizen and Harder 2009). Intensive agricultural practices can compromise pollination services further by reducing nesting opportunities and alternative forage for wild pollinators (Aizen and Harder 2009). This illustrates the complexity of the biological interactions resulting from globalizing intensive agriculture, and unintended consequences of a shift towards homogenous diets that result in large-scale production.

7.2.3 The shift towards more meat-based diets

One of the key dietary shifts in recent times has been the global shift towards a more meat-based diet. There was an average increase in meat consumption of 83 per cent per person between 1961 and 2011, while per capita consumption of starchy roots decreased by 17 per cent and pulses by 28 per cent. Plotting GDP against consumption per capita illustrates that increasing affluence is associated with a shift from staples such as potatoes and pulses to commodities such as meat, milk, and sugar (Alexander et al. 2015). Land used for animal products (that is, both pasture and the land used to cultivate feed), is now nearly twice that attributable to other crops for human consumption combined. By 2006, approximately one-third of arable land was used for feed production (Steinfeld et al. 2006) and about a third of global cereal production is fed to animals (Schader et al. 2015).

As with other aspects of agriculture, due to global trade, the impact of changing diet is not necessarily restricted to the country where the change is located. India is an excellent example of the externalization of environmental impact. Although internally India has relatively low per capita consumption of meat, production is high, and in 2016 India was the world's second largest exporter of beef (FAO 2016).

7.2.3.1 Environmental impact of livestock

The impacts of livestock on the environment are both direct and indirect: directly, where they are reared; indirectly where their feedstocks (predominantly soy and grains) are produced. Through these two routes, intensive livestock production has been linked to water pollution, biodiversity decline, soil loss and degradation, greenhouse gas (GHG) emissions, and climate change.

According to the FAO (Steinfeld et al. 2006), livestock accounts for over 8 per cent of global human water use, most of this being used to irrigate feed crops. The sector is also the largest source of water pollution, which is comprised largely of animal wastes, antibiotics and hormones, chemicals from tanneries, the fertilizers and pesticides used to grow feed, and sediments from eroded pastures. Furthermore, livestock affect cycling in freshwater systems by compacting soil, reducing infiltration, degrading the banks of watercourses, drying floodplains, and lowering the water table. With the loss of soil organic matter through poor grazing and feed production management, productivity declines over time (Gerber et al. 2013).

Machovina et al. (2015) estimated that land cleared for pasture in countries that are home to 'mega diversity' is likely to drive high biodiversity loss and even lead to species extinctions (see Figure 7.1). Consumption of livestock in China and bush meat in Africa is of particular concern with respect to biodiversity loss. Commercialization of hunting for bush meat leads to unsustainable hunting practices that directly reduce the biodiversity of vertebrates, with a tendency to 'hunt down the body size'. In other words, the largest animals are harvested until they are relatively scarce, after which smaller animals are targeted. The logging industry facilitates bush meat hunting by providing access to the forests through transport infrastructure, reducing a hunting trip that may have taken several days to a few hours (Bennett et al. 2002).

7.2.4 Agriculture and GHG emissions

The principal criticism levelled at the livestock industry is that it contributes negatively to climate change through high GHG emissions. This is partly due to enteric emissions from ruminants, i.e. those animals that have a rumen (part of a multichambered stomach). This group includes cattle, sheep, and goats. Ruminants digest tough cellulose-rich plants in a process of microbial fermentation in the rumen, the by-product of which is methane (CH_4) which is emitted as a waste product. CH_4 is

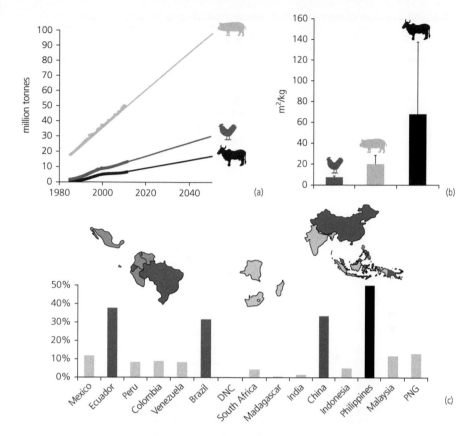

Figure 7.1 Projected increases in area required to produce meat in developing mega-diverse (DMD) countries by 2050. (a) Extrapolating 1985–2012 production data for beef, chicken, and pork in each DMD country to 2050 (data for China shown) multiplied by (b) mean area required to produce livestock biomass provides (c) an estimate of area in each country required to produce livestock in 2050 as a percentage increase beyond total agricultural area in 2012. Agricultural area expansion needs can be met by internal expansion or by agricultural expansion in other countries and importation of feedcrops and/or meat products. The analysis addresses only beef, chicken, and pork. Reprinted from Machovina, B., Feeley K.J, and Ripple W.J. 2015. Biodiversity conservation: The key is reducing meat consumption. Science of the Total Environment 536:419–431, with permission from Elsevier.

approximately twenty times more effective as a greenhouse gas than CO_2. Sources of GHG from livestock production are summarized in Table 7.1.

It has been estimated that livestock production accounts for 14.5 per cent of human-mediated GHG emissions (Gerber et al. 2013). These emissions can be broken down further. By sector, beef and milk production accounts for 41 per cent and 20 per cent, respectively, and pigs and poultry 9 per cent and 8 per cent. By production process, feed production accounts for 45 per cent and processing and enteric fermentation accounts for 49 per cent, manure storage and processing accounts for 10 per cent, with the remainder is associated with processing and transportation (Gerber et al. 2013).

Table 7.1 Summary of some of the GHG emissions relating to livestock production

Supply chain	Activity	GHG	Included	Excluded
UPSTREAM	Feed production	N_2O	Direct and indirect N_2O from: • Application of synthetic N • Application of manure • Direct deposition of manure by grazing and scavenging animals • Crop residue management	• N_2O losses related to changes in C stocks • Biomass burning • Biological fixation • Emissions from non-N fertilizers and lime
		CO_2 N_2O CH_4	• Energy use in field operations • Energy use in feed transport and processing • Fertilizer manufacture • Feed blending • Production of non-crop feedstuff (fishmeal, lime, and synthetic amino acids) • CH_4 from flooded rice cultivation • Land-use change related to soybean cultivation	• Changes in carbon stocks from land use under constant management practices
	Non-feed production	CO_2	• Embedded energy related to manufacture of on-farm buildings and equipment	• Production of cleaning agents, antibiotics, and pharmaceuticals
ANIMAL PRODUTION UNIT	Livestock production	CH_4	• Enteric fermentation • Manure management	
		N_2O	• Direct and indirect N_2O from manure management	
		CO_2	• Direct on-farm energy use for livestock (e.g. cooling, ventilation, and heating)	
DOWNSTREAM	Post-farmgate	CO_2 CH_4 HFCs	• Transport of live animals and products to slaughter and processing plant • Transport of processed products to retail point • Refrigeration during transport and processing • Primary processing of meat into carcasses or meat cuts and eggs • Manufacture of packaging	• On-site wastewater treatment • Emissions from animal waste or avoided emissions from on-site energy generation from waste • Emissions related to slaughter by-products (e.g. rendering material, offal, hides, and skin) • Retail and post-retail energy use • Waste disposal at retail and post-retail stage[a]

[a] Food losses are not included.

Source: From Gerber et al. 2013. http://www.fao.org/3/a-i3437e.pdf.

GHG emissions associated with arable agriculture are the result of the production and use of agricultural inputs, farm machinery, soil disturbance, residue management, and irrigation. Emissions vary between crops, based partly on management. The global demand for cereals has led to large-scale cultivation, and in India, rice is responsible for 37 per cent of GHG emissions (second only to livestock which accounts for 39 per cent), although rice only comprises 9 per cent of the Indian diet. The high emissions are associated with permanently flooded rice, which emits high levels of CH_4 (something that could be mitigated by water management) and relatively high levels of fertilizer application within India. GHG emissions also vary by crop between countries: emissions for cereals are two- to three-fold greater in Europe and North America than they are in India, due to high application of fertilizers in the global north (Vetter et al. 2017).

7.2.5 Loss of traditional crops and varieties

As global diets increasingly converge, local traditional crops and varieties are lost (Maikhuri 1997; Reddy 2007). This results in local food insecurity due to a reliance on imported crops and a loss of on-farm agrobiodiversity and food sovereignty. When traditional crops are abandoned, both the crop and its associated management practices are lost and this can impact on agroecosystems. As an example, traditional intercropping smallholder systems can deliver soil and water protection together with good nutrient and pest management (Bedoussac et al. 2017; Martin-Guay et al. 2018). Of course this is not always the case, but replacement of these locally adapted systems with monocultures and intensive practices is likely to have a detrimental impact on the natural biology. Along with the loss of traditional crops is a loss of knowledge associated with them, leading to subsequent generations being unable to restore a local food system, even where there is a desire to do so (Smith et al. 2018). Thus communities become more reliant on intensive globalized production and the new system becomes entrenched.

The availability of local varieties is also on the decline, as global markets tend to demand homogeneity, resulting in the loss of genetic diversity in crops, yet local varieties are frequently adapted to local variations in climate, soil type, and pests. Their loss means less genetic resources to address potential future challenges such as pest and pathogen outbreaks or climate change (Deb 2017). There are examples of efforts by organizations and individuals working to preserve varieties, and numerous examples of seed saving around the world. See section 8.2.5.

7.3 Global nutrient flows

There are two major impacts that are a consequence of the globalization of agriculture: (1) nutrients are moved around the globe via fertilizers and through traded products; and (2) the movement of these nutrients results in pollution.

Craswell et al. (2004) calculated the nutrient balance of agricultural lands in a range of countries using estimates of the available soil nutrients plus the added organic and inorganic fertilizers, minus the nutrient loss by removal at harvest and by various other loss mechanisms. They determined that the major net importers of nitrogen, phosphorus, and potassium (NPK) nutrients were West Asia, North Africa, and China. Other countries were shown to have a net loss of NPK in agricultural commodities and these were the principal food-exporting countries such as the USA and Australia, and some countries in Latin America. The majority of countries in the study suffered from nutrient depletion (Craswell et al. 2004). Nutrients are also unevenly distributed across farming systems within countries. Although developing countries account for half the global consumption of fertilizers, the majority of these are applied to cereals and cash crops, while subsistence smallholdings receive very little input, so much so that smallholder cropping at current levels is unsustainable in many countries.

Galloway and Cowling (2002) conceptualize nitrogen flows in the context of human activity and illustrate how agriculture interacts with food production (see Figure 7.2).

In countries that import high levels of food and feedstocks, such as Japan, there are challenges with nutrient management which result in pollution and eutrophication of watercourses; agricultural systems are inherently leaky and some are particularly inefficient (Grote et al. 2008). Nutrient import does not necessarily result in improved soils. For example, in Sub-Saharan Africa in the early 2000s, the high quantities of

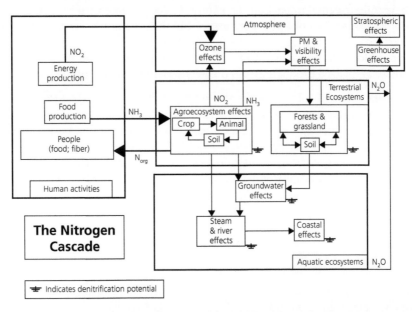

Figure 7.2 Human impacts on nitrogen flows (from Galloway and Cowling 2002). Reprinted with permission from the Royal Swedish Academy of Sciences.

NPK associated with imported food and feed commodities tended to be concentrated in urban areas. These nutrients subsequently resulted in a waste disposal problem rather than addressing deficiencies in rural soils (Craswell et al. 2004).

Livestock production, which is growing with increasing global affluence (see section 7.2.3), is notoriously nutrient inefficient, with an average efficiency of nutrient conversion from feed to animal products of only 10 per cent. Even in well-managed dairy farms, the efficiency ranges between 15 and 25 per cent (van der Hoek 1998). The net result is outflows of nutrients that impact rivers and coastal areas.

The impacts of excess nutrients in the aquatic and marine environments are discussed in Chapter 6, but an excess of nutrients can also have an impact on terrestrial semi-natural habitats, leading to, for example, reductions in biodiversity in grassland (Basto et al. 2015); a negative impact on temperate woodland flora (Pitcairn et al. 1998); and changes to the structure and function of tropical forests (Homeier et al. 2017).

7.4 The spread and impact of exotic pest species linked to agroecosystems

Exotic (or non-native) species are spread via the physical networks that link countries, often by hitching a ride on imported produce, but sometimes by being introduced deliberately or through other means. Many introduced species cannot survive in their new environments, but some are able to establish in new areas and may even aggressively expand their range or may have an ecological or economic impact. These are often defined as 'invasive' species (Lockwood et al. 2013). Invasive species are usually considered agricultural pests if they have adverse effects on production or market access (e.g. post-harvest pests), and many of the pest species within agriculture are exotic. They can result in economic losses by destroying or outcompeting economically or ecologically valued species, in many cases spreading unhindered by their absent host-specific predators.

Many exotic pests are associated with hosts that have also been introduced into new regions. In an analysis of global datasets, Bebber et al. (2014) showed that over 10 per cent of all pests and pathogens have reached more than half the countries containing their hosts. Of these, fungi and oomycetes (fungi-like microorganisms that are often plant pathogens), despite having a narrow host range, are the most widespread and comprise the largest fraction of the most rapidly spreading pests.

Paini et al. (2016) examined the global distribution of 1,300 agricultural invasive pest species over 124 countries, the international trade flows between those countries, and the principal agricultural crops of each. Using these data the authors determined the potential invasion and impact of the study species. While the USA and China were likely to suffer the greatest loss in absolute terms, the most vulnerable countries, relatively, were found to be from Sub-Saharan Africa. The USA and China presented the greatest threat to the rest of the world (given the invasive species they already contain and their trade patterns).

The scale of impact of an invasive species depends on both its ability to disperse and the number of native species likely to be affected by it (although in the case of agriculture, a pest of one key crop species can have a disproportionate economic impact). These are in turn contingent on: species mobility; the presence of suitable prey/hosts; how specialized the species is in terms of diet; the presence or absence of natural predators or disease to control the invasive populations; and tolerance to the environmental conditions. Highly mobile, generalist species can have a high impact on local ecosystems. The cane toad (*Rhinella marina*) is a widely documented example of a species that was imported for a particular task (pest control in sugarcane) but which impacted on native ecosystems worldwide (see Chapter 4, Box 4.4).

The cane toad is a good example of a species used for agricultural management that has impacted on wider ecosystems; in other cases, it may not be clear how pests have been introduced into new regions. The fall armyworm (*Spodoptera frugiperda*), another native of the Americas, was first recorded in Sub-Saharan Africa in 2016, and has since been recorded in twenty-eight countries. The pathway for its arrival although unconfirmed was probably via transport as eggs on imported produce. The fall armyworm is the caterpillar of a noctuid moth. It has a broad diet, feeding on eighty plant species, including maize, sorghum, rice, cotton, and some vegetable crops. It is estimated that 40 per cent of the crops associated with the fall armyworm are economically important. In Africa it has been particularly associated with the destruction of maize, and the adults can move 100 km a night (FAO 2017), meaning that there is a high risk of it spreading to cause destruction across the major crop species. In the early stages of development the caterpillars require very little food and are hard to detect; however, in later stages they are avid feeders and can destroy a maize plant overnight. Pest control is largely by pesticides and, due to low detection rates and the tendency for destruction to take place so quickly, many farmers spray daily. However, appropriate and effective pesticides have yet to be identified and the FAO discourages the indiscriminate use of pesticides. In the long term, the aim is to develop biological control within an integrated pest management approach.

Increased global trade is key driver of pests and diseases in agriculture, and often it is the pathogens that an introduced plant or animal carry, rather than the introduced host, that causes the greatest issues. Two examples of important contemporary fruit diseases, their vectors, and the current approaches to deal with them are outlined in Box 7.1.

Many fungal pathogens are invasive, but their dispersal is not necessarily directly mediated by humans. Asian soybean rust is caused by the fungus *Phakopsora pachyrhizi*. Originating in Asia, it first established in Australia where it caused

Box 7.1 A tale of two pests

In the USA, one of the most vulnerable states for the introduction of pests and diseases is Florida. The region has a mild climate and is a big trade hub. One new invasive species on average arrives every 5 weeks into Miami airport, ensuring that the state has to deal with approximately ten to fifteen new pests per year, and for most of these pests and diseases a multidisciplined approach is needed to try to fight their spread and impact.

Two recent pathogens that have impacted on the agricultural industry in Florida are laurel wilt and citrus greening. They are relatively recent diseases, with laurel wilt first being detected in the USA in 2002 and citrus greening in 2005, and are reported to have had great economic impacts. Both pathogens are carried by insect vectors introduced from Asia, but the biology of the vectors, the pathogens themselves, and the plants' reactions to them differ significantly.

Laurel wilt in avocado

Ambrosia beetles have a symbiotic relationship with fungi—these tree-boring insects cultivate fungi as a source of nutrition. In Florida, the fungi utilized by native ambrosia beetles, such as *Xyleborus bispinatus* (Figure 7.3a), are not detrimental to infected trees, and can actually help break down dead wood; but an introduced species of a closely related beetle, redbay ambrosia beetle (*Xyleborus glabratus*) (Figure 7.3b), brought in a pathogenic fungi *Raffaelea lauricola*. Both the native and introduced species of beetle now carry and spread *R. lauricola* between trees in the laurel family. The laurel family (Lauraceae) in Asia co-evolved with the pathogen and shows no reaction to it, but trees in the USA have no natural immunity. Avocado (*Persea americana*) is a member of the laurel family and is vulnerable to the pathogen. Once a tree is inoculated with the pathogen, it has an autoimmune response (basically an allergic reaction) and is usually dead within 8 weeks. (See Figure 7.4.)

Citrus greening

Citrus greening or citrus huanglongbing (HLB) is carried (in Florida) by the Asian citrus psyllid *Diaphorina citri*, a phloem-feeding insect. The psyllid is a carrier of the

Figure 7.3a *Xyleborus bispinatus*, an ambrosia beetle native to Florida in the USA. Photo: Daniel Carrillo

Figure 7.3b *Xyleborus glabratus* (redbay ambrosia beetle), an Asian ambrosia beetle, introduced into the USA. Photo: Daniel Carrillo

Continued

Box 7.1 (*Continued*)

Figure 7.4 Avocado (*Persea americana*) inoculated with a pathogenic fungus *Raffaelea lauricola* which was introduced with *X. glabratus* and is also carried by other related native ambrosia beetles. Photo: Jonathan H Crane

HLB-causing bacteria *Candidatus liberibacter* species. Both adults and nymphs can acquire bacteria from infected plants. Once carriers, they will retain the bacteria throughout their lives, and are able to transmit them to uninfected plants as they move between hosts. Unlike avocados with laurel wilt, citrus trees do not respond to the presence of *Candidatus* spp. in their systems; instead the bacteria stop the plants from being able to effectively take up certain nutrients, causing yellowing of the leaves, poor fruit development, and eventually plant dieback.

Using biological research for control

For both pests, multiple approaches are being investigated and trialled to try to deal with the issues. Many of these approaches are farmer led and involve management and biosecurity approaches. Alongside these social efforts, pathology, physiology, and entomology are all being investigated.

For example, scientists at the University of Florida are using molecular work to investigate how the beetles interact with and carry the laurel wilt pathogen. Other

Continued

Box 7.1 (*Continued*)

scientists are working on breeding programmes and trying to find resistant root stocks, while still others are investigating why the spread in forest systems is less severe than in agroecosystems. Inoculation of psyllids can be an effective way to prevent them.

It is not just prevention that is being investigated but also control. For example, scientists working on citrus greening have established that thermotherapy (enclosing trees and then steaming them) kills the pathogen but not the tree.

Table 7.2 compares the key aspects of the two pathogens, their vectors, hosts, impacts, and the research methods being employed to control them.

Table 7.2 Comparison between two recent fruit tree diseases and their vectors in Florida, USA

	Laurel wilt	Citrus greening
Scientific name	*Raffaelea lauricola*	*Citrus huanglongbing*
Type of infection (phylum)	Pathogenic fungi (Ascomycota)	Bacteria (Probacteria)
Key agricultural hosts impacted[a]	Avocado (*Persea americana*)	*Citrus* spp., particularly oranges and mandarins
Vector[b] (Class: order: family)	*Xyleborus glabratus* (redbay ambrosia beetle) in regions with native *Persea* (laurel family) trees. Other related ambrosia beetles in commercial avocado groves (Insecta: Coleoptera: Curculionidae)	*Diaphorina citri* (Asian citrus psyllid) (Insecta: Hemiptera: Psyliidae)
Infection location in plant	Vascular tissues—xylem	Phloem sieve tubes
Biological immune response to the infection	Extreme autoimmune disease (allergic reaction). The tree often dies within 8 weeks of inoculation	None. The tree doesn't appear to detect the infection and so doesn't act to respond to it, which enables the infection to spread
External symptoms	Green wilted leaves, followed by desiccation and browning, stem and limb dieback, then death	Yellowing of canopy, dieback of leaves and twigs, reduced size and abnormal shape of fruits. Fruit is bitter tasting. Sometimes death of tree
Standard control methods	Fungicide application prior to outbreaks. Removal and destruction of infested trees	Antibacterial applications. Removal (and destruction) of infected trees. Grafting

Continued

Box 7.1 (*Continued*)

	Laurel wilt	Citrus greening
Integrated biological research strands	Avocado—searching for resistant material via new breeding and rootstocks, and pathology research. Plant physiology studies. Beetle—molecular studies to establish how the beetle interacts with and carries the pathogen. Development of baits. Fungi—pathology research	Investigations into vector–host interactions, and into pathogen–vector interactions. Plant physiology studies. Response to infection. Thermotherapy

[a] Other hosts not listed here are also impacted.
[b] Other vectors not listed here are relevant in other countries.

significant damage in leguminous crops and it has now been recorded in Africa, South America, Hawaii, and along the Gulf coast of the USA. It has been suggested that dispersal was largely by wind. Fungal spores have been recorded at high altitudes and the spores of *P. pachyrhizi* have been found high in the atmosphere. Schneider et al. (2005) suggest that the spores were blown northward into the USA by Hurricane Ivan in 2004.

The rust is known to affect thirty-one different plant species. It spreads quickly and progresses fast, as the spores don't need an open stomata or natural openings in the leaves; sometimes it takes only 6 hours to infect leaves when at a favorable temperature (between 15 and 24°C). Control is largely through fungicides (Delaney et al. 2018). Again, the fungus shows the key strategies for a successful invasive. It is mobile, it infects a range of plants, and it is associated with crop species—which provide plentiful host plants. The growth of monocultures only encourages fungal diseases, and modern agriculture faces some of its greatest challenges from fungal pathogens.

While not all exotic species are damaging and some have a neutral impact or provide tangible benefits to humans (Davis et al. 2011), there are many examples of species that have had an impact on the receiving ecosystem; and while agricultural globalization has been a key driver of the spread of some of these species, agroecosystems are also one of the key ecosystems impacted by them.

7.5 Patterns of land ownership

Patterns of land ownership have changed along with the drive for land expansion and intensification. 'Land grabbing' is the large-scale acquisition of land by both national and transnational companies, governments, and institutions. Although this acquisition of land has gone on throughout history, in particular during periods of colonial expansion, we refer here to a more recent phenomenon where wealthy countries acquire land in countries that are resource rich but economically poor in order to secure food, fuel, and water supplies, and other services. For example, a large amount of land acquisition was prompted by a spike in food prices between 2008 and 2009 (Borras et al. 2011). There are also many instances of private investment pursuing a similar agenda within countries by investing in land. Zoomers (2010) identified seven drivers of large-scale land acquisition by foreign investors, which include, but extend beyond, investment in agricultural production. These are:

- foreign direct investment (FDI) in food production;
- FDI in non-food agricultural products and biofuels;
- development of protected areas, nature reserves, ecotourism, and hideaways;
- Special Economic Zones, large-scale infrastructure works, and urban extensions;
- large-scale tourist complexes;
- retirement and residential migration;
- land purchases by migrants in their countries of origin.

Table 7.3 shows the intention of agricultural land acquisition by percentage of land area acquired in six regions (Nolte et al. 2016).

7.5.1 Impact of land grabbing on agricultural systems

Where foreign investors acquire land for agriculture, it is highly likely that farming practices will be both intensive and not linked to local food security, in order to maximize the return on investment (Kremen et al. 2012). Examples of where this takes place include Argentina, where the provincial government of Rio Negro leased up to 320,000 ha of land to a Chinese government-owned agri-food company to produce soybeans, wheat, and oilseed rape primarily for animal feed (reported in Kremen et al. 2012). As well as the potential impacts on local ecosystems (potential reduction of diversified agriculture for example), land acquisitions can have broader, more social impacts, such as the removal of local autonomy over food production systems, local land disputes, and loss of smallholder farming. Furthermore, it reduces the potential for the development of locally sustainable

Table 7.3 The intention of agricultural land acquisition by percentage of land area acquired based on 2016 data (data from Nolte et al. 2016)

	Americas	Europe	Africa	Asia	Oceania	Global
Food crops	50%	45%	39%	21%	30%	38%
Livestock	16%	17%	3%	1%	11%	8%
Agrofuels	29%	1%	32%	16%	16%	21%
Agriculture (unspecified)	4%	37%	17%	33%	40%	23%
Non-food agricultural commodities	1%	1%	9%	29%	3%	9%

agriculture by directing production towards crops for export and leaves the market vulnerable to price shocks in those countries. The acquisitions increase the commercial pressure on land, leaving some groups particularly vulnerable, such as pastoralists and those dependent on common land such as fishing grounds and forests (De Schutter 2011).

7.6 The influence of technological advances on global agroecosystems

The introduction of new methods and technologies can strongly influence global shifts in agriculture, and this is often led by transnational corporations facilitating the globalization of these new technologies. The development and widespread use of genetically modified crops provide an example of how these scientific and technological developments have combined with other factors to shift global agriculture. Crops are generally modified for (1) pest resistance, (2) herbicide resistance, or (3) nutritional content. An example of pest resistance is Bt cotton, which is modified to resist cotton bollworm. The rise of Bt cotton led to a 212-fold rise in cotton plantings in India between 2002 and 2011, turning it into the world leader for global cotton growing and export (Raman 2017). The adoption of Bt cotton has resulted in a sharp decline in the diversity of cotton varieties grown and 95 per cent of the cotton now grown in India is Bt cotton. However, the crop is still subject to other pest species, and, furthermore, evidence is emerging that pest resistance to the toxin is building (Tabashnik 2015).

An example of a crop modified for herbicide resistance is oilseed rape which is modified to resist glyphosate so that the crops can safely be treated with herbicide without being damaged. The potential for increasing productivity and human health benefits have been, for the most part, welcomed by governments (although Europe remains sceptical), but work to monitor the long-term consequences for the agroecosystem is ongoing. For example, Schütte et al. (2017) looked at herbicide-resistant crops and found that their use reduced biodiversity through the toxicity of glyphosate-based herbicides to a range of aquatic organisms; reducing crop rotation by favouring weed management based on the use of herbicides; the appearance of thirty-four glyphosate-resistant weed species worldwide; and decreasing diversity and abundance of wild plants and arthropods.

Golden rice is an example of a variety developed for nutritional use. It is a patented variety with enhanced beta-carotene to provide vitamin A, and has been designed in particular to address blindness caused by vitamin A deficiency. Although development began in the 1990s and the first field trials were conducted in the early 2000s, golden rice only began to gain approvals for cultivation for human consumption in 2018, so any impact of its development on global agriculture is yet to be seen.

7.7 Next steps

The globalization of agriculture has had far-reaching impacts on the agroecosystem. An increase in population has led to demand for an increase in food production, and this is exacerbated by global and national inequalities in the accessibility of food. High levels of hunger and malnutrition may be in part due to lack of access through multiple aspects of poverty, something that increasing food production will not alone address—yet hunger and malnutrition are still used as the principal rationale for increased food production. Similarly there is a high level of waste in the food system, at all stages of production and consumption (Gustavsson et al. 2011). At the same time, an increase in affluence has driven changes in diets, leading to increased demands for meat in the developing world and for exotic varieties in the more affluent regions of the world. The global market for food has resulted in the inequitable distribution of nutrients, land-use change, the simplification and specialization of agroecosystems, a reduction in biodiversity, an increase in the use of fertilizers and pesticides, and high GHG emissions. The challenge is to mitigate the negative impacts while enabling food security, recognizing that solutions cross the boundaries of disciplines and this requires a collaborative approach. Chapter 8 outlines some approaches to meet these challenges.

8

Conservation and Sustainable Management of Agroecosystems

8.1 Introduction

This chapter focuses on the role of farm management in biodiversity conservation (and the potential conflicts and synergies between them). There is a balance to be made between crop production and management of wildlife and diversity in farming systems. Land may be 'spared', where a portion of the landscape is used for intensive food production and another portion is retained for wildlife, or it may be shared, where diverse cropping and semi-natural habitats share space. Most farms fall within a gradient of production, from high-input intensive systems to diverse low-input poly-cultures. In this chapter we focus on the land sharing approach, where biodiversity and semi-natural habitats are combined, to various degrees, with the shared aim of food production, biodiversity conservation, and ecosystem service provision. Clearly there will be trade-offs between the three, as well as interactions between them, and these will need to be weighed up by those making land management decisions.

There are two key approaches to combining agricultural production with biodiversity conservation and ecosystem service provision: (1) introducing diversity into the cropped areas, either through crop diversity and poly-cultures (e.g. intercropping) or through integrating non-crop habitats into the crop (e.g. living mulches); and (2) managing semi-natural habitats within the farm boundaries, but in areas that are separated from the crops (e.g. introducing flowery field margins). Additional innovative approaches to achieve conservation and sustainable management of agroecosystem can be implemented, and will influence future agroecosystems.

8.2 Reintroducing diversity into agroecosystems

Reintroducing diversity into agroecosystems can help enhance the networks within a system; this may in turn increase productivity (see Box 8.1). Some examples of different methods to reintroduce diversity into agroecosystems are outlined in the next sections.

The Biology of Agroecosystems. Nicola P. Randall and Barbara Smith, Oxford University Press (2020).
© Nicola P. Randall and Barbara Smith 2020. DOI: 10.1093/oso/9780198737520.001.0001

Box 8.1 Improving understanding of agroecological systems using a network approach (above- and below-ground networks)

The relationship between above- and below-ground networks of organisms underpins the structure of terrestrial ecosystems. Plants are a link between the two, making connections between above-ground herbivores, pathogens, mutualistic symbionts, and antagonists to both the soil and rhizosphere (root space) communities (Ramirez et al. 2018). It has been demonstrated that communities and processes are intrinsically linked and that there are feedbacks between the two subsystems which are important for ecosystem functioning.

Soil organisms below ground have low mobility and longer survival rates than those above ground, and therefore plants may have longer-lasting legacy effects below ground than above (van der Putten et al. 2009). Soil biota is highly diverse and the microbial community may comprise thousands of species with complex interactions.

One way to understand the relationship within those complex systems and the relationship between above- and below-ground communities is to use a network approach. Many studies have used simplified systems to understand the relationship between the subsystems, which has enabled a good understanding of specific mechanisms. However, as there is increased interest in understanding the whole system, mapping linkages can clarify dependencies and antagonisms (Ramirez et al. 2018). Plate 6 demonstrates how a network approach can help explore these relationships. In agroecosystems these relationships are critical for crop production, and this whole-system approach enables researchers to make predictions about below-ground changes that can impact on above-ground productivity.

8.2.1 Crop rotations

One way of diversifying systems is to increase the number of crops in the cropping rotation. Rotations are a useful tool in protecting crops from pests and diseases, and in maintaining or improving soil fertility. Growing the same crop in the same soil over a period of years results in a depletion of nutrients and enables the build-up of pathogens and pest populations. Rotating crops simply means ensuring that the same crop is not sown in the same field in consecutive years, but that each crop is followed by another, preferably from a different family. Crop rotations are usually formalized in a sequence. Simple rotations might be comprised of one cereal and an oilseed. More complex rotations might, for example, comprise more than one cereal, row crops such as vegetables, and a green manure containing fertility-building legumes. Rotation can break disease and pests cycles, especially when alternating between cereals and broadleaf crops. In some cases, a single year of an alternate crop will be sufficient to break a disease cycle, although for others, where the disease is more persistent, it will be necessary to devise more complex rotations. In Africa, crop rotation has been a useful tool to reduce populations of *Striga*, or witchweed, which is a hemiparasitic plant that is a particular problem in maize. Experiments with rotations demonstrated that rotating crops reduced

Striga seed numbers in the soil (Oswald and Ransom 2001) and showed that a good sequence to reduce the *Striga* weedbank is maize planted after a two-season rotation that included pigeon pea (Oswald and Ransom 2001). In another example it was shown that including a pulse crop can stimulate beneficial organisms and reduce the severity of root rot in cereals (Krupinsky et al. 2002).

Another key benefit of using rotations is soil and fertility building. Magdoff and Van Es (2009) described a long-term experimental trial in Missouri that compared continuous maize with various rotations and a continuous timothy grassland. After 60 years the maize plots had lost 56 per cent of the soil when compared with the retained soil in the grassland plot. In comparison, a 6-year rotation consisting of corn, oats, wheat, clover, and 2 years of timothy led to only 30 per cent loss. Similarly, work in Argentina showed that increasing rotational diversity from two crops to four increased the metabolic activity and diversity of the soil microbial community and abundance of the soil microbiota, which can have beneficial effects on soil aggregation and nutrient cycling (D'Acunto et al. 2018). There is some evidence that crop yield can be enhanced using rotations. It has been reported that yields of crops grown in rotations, in the absence of inputs, were between 30 and 100 per cent higher than those of crops grown in monoculture (Smith et al. 2008) and other studies have reported similar, if not such dramatic effects (Lund et al. 1993). Including livestock in rotations can have multiple benefits, including the delivery of organic matter and nutrients to the soil via dung, and this is discussed below.

8.2.2 Intercropping and polycultures

A further level of complexity can be introduced into fields using intercropping, which is the practice of growing two or more crops close together, either as an unstructured mixture or in conventional rows. It is an ancient traditional practice that is still practised over much of the world in smallholder systems and in some regions it remains a dominant form of agriculture.

In Latin American smallholder farmer systems, 70–90 per cent of beans are grown with maize, potatoes, and other crops. In Africa, 98 per cent of cowpeas are intercropped. Across the tropics, the proportion of cropped land that is intercropped varies from 17 per cent in India to 94 per cent in Malawi (Brooker et al. 2015 and references therein). In Europe, intercropping has been used extensively in organic systems, and increasingly researchers are investigating 'novel' crop combinations in conventional agricultural environments, combining, for example, cereals with legumes (Bedoussac et al. 2017). Intercropping has been shown to increase yield, improve soil quality, and increase soil carbon sequestration (Cong et al. 2015), and reduce run-off and soil erosion (Zougmore et al. 2000); moreover, biomass production in species-diverse systems has been shown to be, on average, 1.7 times higher than in monocultures (Cardinale et al. 2007).

Brooker et al. (2015) summarize how higher yield gains are mediated through facilitation, resource sharing, and niche complementarity (Figure 8.1). Facilitation is the process whereby the combination of plants increases the resources required to the common

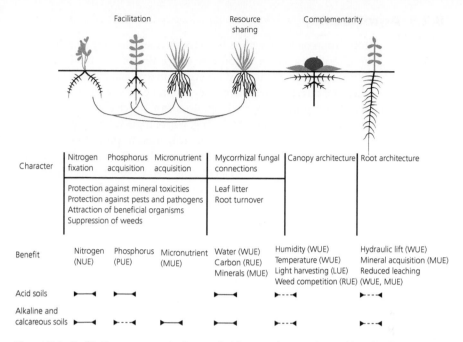

Figure 8.1 Facilitation, resource sharing, and niche complementarity enable polyculture systems to yield more than their corresponding monocultures. Certain facilitative interactions can be associated with particular soil types (either acid soils or alkaline and calcareous soils), and when present can be either strong (solid lines) or weak (dashed lines). Reproduced from Brooker et al. (2015)

pool. As examples, nutrient availability can be increased via nitrogen fixation, the combination of plants can increase weed suppression, or one species might encourage the beneficial organisms that control pests in another species in the system. Resources can be shared via common mycorrhizal fungal networks or resource turnover making nutrients more available. Niche complementarity allows optimal exploitation of light and soil resources, and it can be observed in species with contrasting architectures (such as short and tall shoot architectures, or shallow and deep root architectures).

However, yield increases are not consistently greater in intercropped systems, and tend to be better in comparisons against cereal crops rather than against leguminous crops. In a meta-analysis of over 900 different observations, Martin-Guay et al. (2018) showed that when compared to the same area of land managed in monoculture, intercrops produced 38 per cent more gross energy and 33 per cent more gross income on average, while using 23 per cent less land. However, intercropping is not consistently superior in all cases under all conditions, and where there are high inputs, intensively managed monocultures can produce higher yields (Zhang 2007a).

Another kind of intercrop is the 'living mulch' which undersown with cash crops. Their purpose is to suppress weeds and prevent soil erosion. Legumes are frequently sown as living mulches for the nitrogen fixing abilities. However, the competition between the crop and the living mulch may result in lower crop yields.

8.2.3 Agroforestry

Agroforestry (intercropping trees with other crops) is another type of diversified agriculture. The global area under agroforestry is estimated to be 1 billion ha (Brooker et al. 2015). According to the FAO there are five principal types of agroforestry. Three of these are directly productive: alley cropping, where wide rows of trees are interspersed with companion crops; silvopastoralism, which combines trees with grazing or forage production; and forest farming, where high-value crops are grown under the forest canopy. The other two are primarily used to protect crops or livestock: riparian forest buffers, where rows of trees are planted alongside rivers and streams to inhibit sediment and pollutants entering watercourses; and windbreaks, which are sown to protect soil, crops, and livestock from wind and snow.

In addition to their primary function, agroforestry systems can increase agricultural productivity, improve soil fertility, control erosion, conserve biodiversity, and diversify income for households and communities (Feliciano et al. 2018). There has been increased interest in agroforestry systems due to their ability to sequester carbon in both above- and below-ground biomass (Ramachandran Nair et al. 2009). In a meta-analysis, Feliciano et al. (2018) demonstrated that carbon sequestration will vary according to plant characteristics (species, age, biodiversity, and tree density), farming system characteristics (structure, function, stability), and management factors (tillage, fertilization, residues, holding size, and harvesting regime). These, coupled with local climatic conditions and soil characteristics, mean that spatial and temporal variability will impact on the potential of any given agroforestry plot to sequester carbon.

Thorlakson and Neufeldt (2012) trialled agroforestry systems in Kenya. In this case, community groups were provided with 200–300 tree seedlings, agroforestry training, staff support (for up to 1 year), and tools for tree nursery establishment. The tree species included in the trial were *Acacia mellifera*, *Acacia polyacantha*, *Albizia coriaria*, *Calliandra calothyrsus*, *Casuarina equisetifolia*, *Cordia abyssinica*, *Faidherbia albida*, *Gliricidia sepium*, *Grevillea robusta*, *Markhamia lutea*, *Senna siamea*, and *Warburgia ugandensis*. Control groups were established in neighbouring communities. Both agroforestry and experimental groups were exposed to drought and flood conditions, which were present within the region during the study period. Productivity was greater among the agroforestry group (many farmers chose nitrogen-fixing trees); their income improved due to sales of fuel wood, timber, fruit, and seedlings, and savings in food purchases enabled by their increased productivity. Soil erosion was reduced. Finally, farmers were able to offer alternative crops despite the extreme weather conditions.

In Europe there are examples of traditional agroforestry systems that integrate livestock, such as pig production in the cork oak forests of Andalucia (Campos et al. 2018). Increasingly, agroforestry is being considered as a solution to environmental challenges; in the UK, agroforestry is being explored as a tool for flood management and in climate-smart agriculture (Burgess 2017). In some regions, agroforestry is embedded in production systems. Chakras in Ecuador are a good example of this (Box 8.2).

Box 8.2 Forest gardens in a global market

The market for products with health-giving properties is growing. One new product, a tea, is harvested from the leaves of *Ilex guayusa* (guayusa), a native tree of the western Amazon region grown by indigenous farmers. The tea is said to have energy-giving properties and it has been successfully commercialized and marketed outside the Amazon. Guayusa is an increasingly popular new crop for farmers in Ecuador and is grown in 'Chakras'. Chakras are indigenous agroforestry plots that are passively managed and provide food security for local populations; they are essentially modified forests, which to the untrained eye can resemble wild tropical forest. A Chakra contains both perennial crops (such as cocoa, guava, citrus, coffee, plantains, a variety of palms, medicinal plants, and timber trees) and annual crops (such as pineapple, cassava, sweet potato, corn, peanuts, pulses, and tubers) and may contain some fifty edible or usable species.

The advantages of growing guayusa are that it can be incorporated into the existing Chakra system and provide an alternative to cocoa and coffee, or at least reduce a farmer's reliance on them; the trees are quick growing, so that leaves can be harvested in the first year; and they provide shade and potentially support biodiversity, although there is a risk that farmers will plant in dense stands, reducing species richness and increasing the potential for pests and diseases to spread. A further risk is due to poor genetic diversity in the crop, as most plants are clones, leaving the crop vulnerable to pest or pathogen outbreaks. With good management, the forest gardens offer an opportunity for species-rich habitats to survive, while providing income for farmers (Krause and Nesse 2017).

8.2.4 Alternative grassland management

8.2.4.1 Sward diversification

Pasture is frequently managed as a monoculture (usually of rye grass), with high levels of nitrogen application. However, Schader et al. (2015) noted that grasslands can contain large carbon stocks and provide many ecosystem functions which would be lost were the land converted to croplands, yet the composition and management of pasture has an impact on the environmental impact of grasslands. Heavily fertilized monocultures of grasses such as rye grass, together with animal waste, can have a negative impact through increasing nutrient run-off into watercourses, and over-grazing leads to soil compaction and erosion.

Diversification of plant species in pasture is an opportunity for sustainable management. Diverse swards can provide a more nutritious diet for livestock that requires less nitrogen inputs and has the added benefit of increasing biodiversity and resources for higher trophic levels (Beetz and Rinehart 2006). Careful selection of species for pasture can enhance these efforts; for example, *Medicago sativa* (lucerne), *Onobrychis viciifolia* (sainfoin), and *Lotus corniculatus* (bird's-foot trefoil) are common meadow plants that are known to have secondary compounds that reduce methane production (Beauchemin et al. 2008) and also support wider ecosystem services. Even modest enhancement of pasture by the addition of forbs and legumes can increase pollinator diversity and abundance (Orford et al. 2016). However, poor dispersal and a need for gaps and optimum sward height to support the germination and growth of desirable species (Bischoff 2002) mean that diversification of pasture needs careful management.

8.2.4.2 Rotation grazing systems

Another way to improve ecosystem services in grasslands is to change the grazing regime of associated livestock. This may involve changes to the species or breed of livestock or to the ways in which the land is used to support them.

In natural grasslands, wild grazing species tend to roam from one place to another in herds, meaning that the grasslands alternate between periods of low grazing and trampling intensity to (often comparatively shorter) periods when this is very intense as animals move through a region. This contrasts with many non-nomadic agricultural livestock systems, where grazing is continuous (livestock are grazed on one area for either a large part or all of a growing season, or even year-round (Hawkins 2017). Grazing systems that are designed to mimic natural herd migrations are growing in popularity in many regions. Variations of these include rotational grazing (moving livestock from pasture to pasture), mob grazing (where high densities of stock are grazed for short periods), and holistic planned grazing (HPG) (rotational grazing, where adaptive management is used to guide the timing of rotations). Advocates of these methods claim that they can improve both productivity and environmental quality when compared to continuous grazing, but the scientific evidence does not necessarily confirm this, with individual study results highly variable. It may be that this is related to the type of grassland and external management factors, with one meta-analysis finding that HPG was more likely to have a positive impact on productivity in regions with higher precipitation levels (Hawkins 2017).

8.2.5 Genetic diversity

A less immediately visible way of diversifying systems is to consider genetic diversity within crops. The importance of maintaining genetic variety within crops is based on the assumption that there are beneficial traits for pest and disease resistance, or traits that confer resilience to particular environmental conditions, that are held in traditional varieties. Conserving varieties of crops can be equated to maintaining a bank or library of traits, which can be incorporated into commercial crops where necessary. Using well-established breeding techniques, by backcrossing commercial varieties with wild species with desirable traits, it is possible to create varieties that remain high yielding.

Developing a new variety in this way requires extensive trials to ensure that the target traits are incorporated and stable, although this is simpler if the genes have been sequenced and the genome can be analyzed with molecular techniques. It is estimated that it takes 5 years to develop a new variety (Gruber 2017). In the USA, there is extensive work being done on tomatoes to improve nutritional content; for example, a team at Oregon State University in Corvallis has been breeding tomatoes with increased amounts of anthocyanins (antioxidants). By backcrossing with wild relatives, they have bred a variety that has between 10 and 30 mg of anthocyanin per 100 g, whereas regular tomatoes contain none. However, the fruit are a deep purple rather than the usual red, and commercial growers will therefore have to make the decision as to whether consumers will accept purple tomatoes (Gruber 2017)!

Libraries or gene banks are being established to conserve wild relatives of crops and traditional varieties, as there is concern that we are losing them; more than 70 per cent of wild relatives are considered under threat (Gruber 2017). Examples of gene banks include the Svalbard Global Seed Vault in Norway and the Millennium Bank at Kew Gardens in the UK. In-situ conservation of rice varieties was adopted in China in 1997, since when the number of traditional rice varieties in cultivation has increased dramatically and now includes some varieties that were formerly locally extinct (Zhu et al. 2003); in India the Vrihi seed bank held 1,400 accessions of rice by 2019; in the UK, the Heritage Seed Library holds a stock of old vegetable and grain varieties that are no longer licensed for sale.

These living gene banks have the potential to protect food production in the future, and can sit well alongside high-tech solutions such as gene editing and crop breeding.

8.2.6 Reintegrating cropping systems with livestock

Increasingly, cropping systems and livestock systems have been spatially separated. Changing land-use patterns by integrating cropping systems and livestock would result in: (1) a diversified landscape with structured habitat mosaic; (2) spatial and temporal interactions between different land-use systems to better regulate biogeochemical cycles; and (3) increased resilience of the system to socio-economic and climate change (Lemaire et al. 2014).

8.2.7 Regionally appropriate breeds

Allied to ideas of genetic diversity in crops is genetic diversity in livestock. In recent times, livestock breeding programmes have narrowed the genetic stock and this has led to the loss of regionally adapted breeds. Although high-yielding breeds have seen production increase in some areas, this has not been a universally successful approach, particularly in places where non-local breeds have been imported.

Cattle farming provides a prime example of the challenges associated with importing non-local breeds. Globally there has been a move to import Holsteins into tropical countries such as India and Sub-Saharan Africa. Holstein cattle originated in the Netherlands and produce notably high milk yields. With this in mind, successive governments and NGOs encouraged farmers in developing countries to raise imported Holsteins in a well-meaning attempt to address food insecurity. Although bred to be high yielding in temperate climates, Holsteins and similar varieties had poor resistance to the tropical diseases in these new locations and failed to reach their expected potential. In some cases (for example trypanosomiasis, a disease transmitted by the tsetse fly in West Africa), the cost of drugs has outweighed any increase in income because local breeds had inherent resistance. Furthermore, these expensive but delicate introduced breeds were less likely to be let out to pasture, and farmers were trapped in a cycle of providing costly and often imported feed (Eisler et al. 2014).

Using local breeds and new breeding programmes enables farmers to engage in more sustainable farming practices, although more work needs to be done to persuade farmers of the potential value of local breeds (Eisler et al. 2014). The integration of locally adapted breeds in intercropping or agroforestry systems has potential to contribute positively to sustainable management in agriculture.

8.2.8 Diversification of food systems

Another approach to diversifying agriculture is to consider it from the perspective of what we consume. There is an emerging counterbalance to dietary/landscape homogenization. Across the world there are movements to encourage the restoration of more traditional and diverse diets. This is exemplified by the Slow Food Movement which began in Italy in 1989 and now has branches in 150 countries. The aim of the movement is to promote local foods and traditional diets and food production and is a conscious move away from globalization and industrial agriculture. Allied movements proliferate. For example, in India there are now forty seed banks dedicated to collecting and preserving the seeds of traditional foods such as millets and diverse varieties of rice, some of these such as the Vrihi seed bank (see section 8.2.5) make their seeds available to farmers. On the global market there are just a handful of rice varieties for sale, but the Centre for Interdisciplinary Studies (CIS) is a seed-saving organization that now holds at least 600 varieties of rice from West Bengal alone.

The movements are relatively small and are exploiting alternative, or niche, markets, but they are making a positive impact on conserving diversity in the agroecosystem. Increasingly there are renewed calls for diversified farming systems explicitly aimed at maintaining ecosystem services such as soil fertility, pest and disease control, water efficiency, and pollination (Kremen et al. 2012).

8.2.9 Creating or enhancing non-productive areas on farmland

Non-productive areas on farms can be used to increase biodiversity and provide ecosystem services. For example, it is well known that floral strips can increase services such as pollination and pest control by providing resources for beneficial insects that need floral resources as part of their lifecycle (Feltham et al. 2015; Westphal et al. 2015). Riparian buffer strips can act as 'catch' crops, intercepting pollutants and protecting water and air quality (Blanco-Canqui et al. 2015), and strips of seed-bearing plants have been shown to increase the abundance of farmland birds (Stoate et al. 2004). Existing features such as hedgerows can be improved and better managed to provide refuge and food for a wide range of farmland wildlife by providing food and shelter for insects, birds, and mammals. Finally, increasingly, there is work investigating whether the integration of areas dedicated to biodiversity in agricultural landscapes can enhance the wellbeing of the people that work, and play, in those areas. Although the research is in its infancy (see Saxby et al. 2018 as an example), there are many examples of work demonstrating that exposure to semi-natural habitats can improve wellbeing (e.g. Wood et al. 2018).

While there are general principles that can be used to design non-productive areas, the site-specific context will have a big impact on their outcome. Microclimate, soil type, slope, and moisture all determine which species will flourish and which will struggle. Land managers use their knowledge of their land to determine where to site particular non-crop habitats. In an ideal scenario, the least-productive areas of the farm will be suitable for encouraging wildlife. Field corners and areas that are difficult to harvest with machinery can be usefully sown with wild plants or can be left to colonize naturally. Providing there are no pernicious weeds present that are likely dominate, these areas can contribute to biodiversity on farm.

Another approach that makes use of less-productive areas is the establishment of 'conservation headlands', where sprays of pesticides and fertilizers are minimized or eliminated in the outermost 6 m of cereal fields. Research has shown that this simple approach can increase arable flora and provide insect food and nesting space for farmland birds (Vickery et al. 2009). If left uncut over winter, they can additionally provide seed for birds.

The benefits of semi-natural and low-input areas are best seen when used in combination in a coherent design. For example, a well-managed hedgerow adjacent to a conservation headland will maximize the value of both by providing complementary resources for wildlife. Connecting semi-natural habitats can provide corridors along which animals can travel and plants can spread. There will inevitably be trade-offs between priorities for production and wildlife, but a strategic approach to landscape design will maximize the environmental benefits.

Land abandonment is a problem in many countries. Theoretically this may offer some positive opportunities for improved biodiversity as wild areas are considered to be rich in wildlife. However, the value of farming systems can be underestimated. For example, extensive grazing systems can maintain species-rich meadows and loss of these low-impact farming systems can bring about concomitant loss of biodiversity. Furthermore, smallholder landscapes can be diverse. A case study in the mountainous region of the Barcelona metropolitan region (Spain) showed that a move from smallholder farming with numerous vineyards to forest cover resulted in reduced biodiversity (Otero et al. 2015). The authors suggested that the biodiversity value of a heterogeneous smallholder-dominated landscape is frequently underappreciated by those advocating the benefits of land abandonment (or rewilding).

8.3 Managing pests and nutrients sustainably

Pests and nutrient availability are two principal limiting factors in food production. There has been a reliance on synthetic inputs to manage both and it is a challenge to do so sustainably. In organic systems, synthetic inputs and pesticides are excluded; however, in conventional systems, mixed approaches have been developed and novel technologies employed, the aim being sustainable farming alongside high yields.

8.3.1 Integrated pest management

The FAO defines integrated pest management (IPM) as 'the careful consideration of all available pest control techniques and subsequent integration of appropriate measures that discourage the development of pest populations and keep pesticides and other interventions to levels that are economically justified and reduce or minimize risks to human health and the environment. IPM emphasizes the growth of a healthy crop with the least possible disruption to agro-ecosystems and encourages natural pest control mechanisms.' Within this broad definition there are a number of established approaches.

8.3.1.1 Classical biological control

Classical biological control (or biocontrol) is the active introduction and release of biological control *agents* such as predators, parasitoids, pathogens, and competitors to control pests (both plant and animal). It is used to control crop pests and weeds, especially invasive species, which, being non-native, often lack predators or disease to control them.

There is tight regulation around the release of non-local species due to the negative impacts of early introductions. For example, in 1906, populations of the European fly (*Compsilura*) were released in North American forests to control a serious tree pest, the gypsy moth (*Lymantria*). Releases continued until the mid-1980s. However, it is now known that the fly attacks more than 180 species of Native American insects and has been influential in the decline of four species of silk moth. Yet there are also many successful examples. *Opuntia stricta* (prickly pear) was introduced into Australia in the early nineteenth century and quickly became invasive, eventually infesting 30 million ha. After a global survey, the prickly pear herbivore the moth *Cactoblastis cactorum* was introduced; within a decade, an estimated 1.5 billion tonnes had been cleared.

Classical biocontrol continues to be developed for highly invasive species. Both Europe and the USA have witnessed the seemingly unstoppable spread of Japanese knotweed (*Fallopia japonica*). In response, a fungal pathogen (Kurose et al. 2012) and a *Pysillid* bug (Clements et al. 2016), both sourced from Japan, have been identified as potential control agents and are undergoing rigorous biosecurity trials. The most important criterion is specificity, to ensure that non-target organisms are not affected.

8.3.1.2 Augmentation

Less risky is the augmentation of local native populations of natural enemies, whereby parasites or predators are either periodically released to increase their numbers or the supply of their food resources is increased, in order to ensure that adequate numbers of parasites and predators are present to provide the desired level of pest control (Ridgeway et al. 1977). A well-known success story is the case of using entomopathogenic nematodes (EPN) for management of the root weevil, *Diaprepes abbreviates*, in Florida citrus groves. In this case, the nematodes are released in great numbers as a biopesticide. However, studies raise the concern that there could be some negative impact on endemic nematodes and suggest that habitat manipulation might be a more sustainable approach (Stuart et al. 2008).

8.3.1.3 Conservation biological control

Managing habitats on farmland to deliver IPM is a low-impact, risk-free approach. Complex, permanent habitats such as hedges, grassland, and woodland patches can increase the abundance of pest natural enemies in the landscape, as can temporary habitats such as flowery field margins, grassland strips, and beetle banks. These habitats provide shelter and food resources (including alternative prey) for the organisms that control pests in the cropped fields (Hassan 2016).

With this approach there are additional benefits, including non-target biodiversity and other ecosystem services. It requires a landscape approach because most natural enemies experience their habitat at larger scales than a single field. Important criteria for successful pest control are the distance of the habitat from the field (for example, many invertebrate natural enemies will not travel further than 80 m away from a field margin) and the diversity of the surrounding landscape. However, there is no doubt that this method is less predictable and less amenable to targeting than an approach such as large-scale release. This can be ameliorated by encouraging species diversity (to buffer against spatiotemporal disturbances) and connectivity in the landscape, ensuring that natural enemies can disperse (Tscharntke et al. 2008).

8.3.2 IPM in protected cropping systems

IPM has been very successfully used in protected cropping systems (greenhouses and polytunnels), which enable more control, and in many countries it has become the principal form of pest control in protected cropping. In Spain, for example, the tomato and sweet pepper industries rely heavily on IPM.

In some cases, banker plants are used in greenhouses. Banker plants are plants introduced to protected cropping systems as long-lasting rearing units for beneficials—the plants are either infested with herbivores or supply other food items for beneficials, such as pollen. This is a low-cost technique which ensures a fresh supply of beneficials in the system and works well in an IPM programme (Huang et al. 2011).

8.3.3 Sustainable nutrient management

Sustaining nutrients on farmland is essential for continued productivity, and research in the UK has shown that fertilizers (either artificial or manures) increase yields by two to three times. Over-application has led to nutrient surpluses in many farming systems, however, and these surpluses can leach out of the agroecosystems themselves and into surrounding ecosystems. There are multiple ways to address nutrient issues. These largely focus around soil management and include minimizing and optimizing inputs (i.e. only applying nutrients when most needed, something that has become easier to assess with modern technologies such as precision farming); utilizing legumes within crop rotations to fix nitrogen more readily within soils; maintaining soil structure through management to reduce compaction; introducing or maintaining organic matter; and maintaining ground cover wherever possible (Goulding et al. 2008). Many of these processes are included as more general integrated approaches to agriculture.

8.4 Alternative agricultural approaches

There are alternative approaches to agriculture that take a more holistic approach to farming than conventional farmers and that some consider inherently sustainable.

8.4.1 The organic approach

Organic farming is a method of crop and livestock production that does not use pesticides, fertilizers, genetically modified organisms, antibiotics, or growth hormones. This applies to both crops and livestock; all inputs such as seeds, compost, and animal feed must comply with the same criteria. Globally, in 2016, 57.8 million ha were recorded as being under organic production, which is 1.2 per cent of the total agricultural land. This represents an increase of 46.8 million ha since 2011; 178 countries use some proportion of organic production, 87 of which have organic regulations. The global market is growing and the market share is increasing; the global market is currently worth 80 billion Euros. The top three countries in terms of market share are the USA (38.9 per cent), Germany (9.6 per cent), and France (6.7 per cent) (Willer et al. 2018).

Compared to conventional farming, organic systems deliver equally (or more) nutritious foods that contain less (or no) pesticide residues (Reganold and Wachter 2016). Although yields may be lower on average, this may not be true in every case. For example, in a study of conventional versus organic cotton systems in India (each with a rotation of cotton and wheat), yields in the organic system were lower in the first year (by 29 per cent for cotton and 27 per cent for wheat) but were similar to the conventional system in the second crop cycle, where yields in the conventional system reduced. Organic soybean (a nitrogen-fixing legume) yields were marginally lower than conventional yields, but soybean gross margin was significantly higher in the organic system (+11 per cent) across four harvest years compared to the conventional systems. This suggests that organic soybean production with cotton could be a viable economic option for smallholder farmers in semi-arid parts of India (Forster et al. 2013).

A number of papers report that organic systems support more biodiversity than conventional farms. This has been demonstrated for many groups of flora and fauna, including arable plants (Shepherd et al. 2003), butterflies (Rundlöf and Smith 2006), bees (Holzschuh et al. 2007), and birds (Chamberlain et al. 2010). There are also reported trade-offs, notably concerning yield (Gabriel et al. 2010). However, there is a debate over how fair the comparisons are, as like is not always compared to like. Wheat yields from a conventional arable farm compared with wheat grown on an organic mixed farm may not be a level playing field. One could also argue that the greater biodiversity on organic farms is partly due to organic farms usually being mixed and that this introduces a bias against conventional systems. A challenge in understanding production versus biodiversity in farmland is the lack of whole-farm modeling which tracks total production. However, as a generalization, organic farming is good for biodiversity. The lack of insecticides and herbicides in organic farming is at least partly responsible, but organic farms are also likely to be more structurally complex, which promotes biodiversity (Norton et al. 2009).

Without pesticides, organic farmers have had to develop alternative methods to deal with insect pests and weeds, and many of these have migrated into conventional systems.

8.4.1.1 Weed management

Weed management is an approach that favours the crop over the weeds. The aim of weed management in organic systems is not to eradicate weeds entirely but to reduce the numbers to a tolerable number (another reason why the system may support biodiversity). This makes organic management knowledge intensive. Rotations can be positively managed to influence weed control, and crop planning is very important in organic systems. There are no prescriptions and individual farms vary, but, as a general rule, autumn and spring crops are alternated as are annual and perennial crops; and dense crops (such as oats) that shade out weeds, are rotated with open crops (such as maize) which encourage weeds. Rotating cattle, sheep, and pigs is used to ensure that no one species dominates. Physical removal or burying of weed seeds using tillage is a common approach and farm hygiene is also emphasized in organic systems, i.e. being careful not to move weed seeds around the farm with dirty machinery and not introducing weed seeds into the system in imported composts (Davies et al. 2008). Because tillage is such an important weed management method in organic cropping systems, minimum tillage and other conservation tillage practices are often not an option.

8.4.1.2 Insect pests and pathogens

Insect pest infestation is discouraged in organic systems by using pest-resistant varieties and rotations to reduce pest build-up. If pests do build-up, biological control mechanisms can be used, either by the release of natural enemies or through conservation biological control, as described above. In the case of some pathogens that are very harmful to crops, simple inorganic chemicals are approved for organic systems, such as copper against potato blight. However, the discovery of negative environmental impacts means that these are being phased out. Pests and pathogens remain a challenge (Davies 2012).

8.4.2 Biodynamic agriculture

Biodynamic agriculture is similar to organic farming in that there is no use of synthetic inputs, but it goes further by conceptualizing the farm as a system of ecologically interrelated tasks rather than an assembly of separate crops and livestock. Biodynamic agriculture aims for a closed loop system (where nutrients are neither imported nor exported but recycled on site) and to support local production and distribution systems and traditional/local breeds and varieties. Although these concepts are beginning to be incorporated into more conventional organic systems, they were revolutionary at the time they were developed in the 1920s by Rudolph Steiner. It was the first explicitly organic system.

Unlike conventional approaches, there is a spiritual and philosophical dimension that diverges sharply from most farming management systems; nevertheless, biodynamic agriculture was recorded as being practised in sixty countries by Paull (2016), although just under half of the recorded 162,000 ha under biodynamic cultivation was in Germany, its country of origin.

8.4.3 Integrated approaches

The commonality between many alternative agriculture approaches is that they aim to offer a more integrated approach to management of agroecosystems. This may or may not involve the use of artificial fertilizers or of pesticides, and it may or may not include pioneering technologies, but integrated approaches usually aim to try to work, at least to a point, with natural systems, only intervening where necessary (e.g. monitoring pests and applying management accordingly, rather than applying pesticides indiscriminately). Agroecology (see Chapter 9, section 9.4.1.2), integrated farming systems (IFM), and circular agriculture are loosely linked terms that may be used to describe these more integrated approaches, and they may apply within single farms, within landscapes, or within policy approaches to agriculture. An example of a more integrated approach by an individual farming in western Europe is outlined in Box 8.3.

Box 8.3 An integrated approach to agroecosystem management: a UK case study

Typically in the UK, the trend has been to become more specialized and produce fewer products from one farm. Greenacres Farm is a mixed organic farm in Shropshire, and so bucks the trend. Sheep and cattle are farmed for lamb and beef products alongside multiple different crops and some agroforestry. The aim is for the farm to be a complete system, and it runs on a 5-year rotation. This starts with a clover ley to support grazing livestock and provide nitrogen to benefit crops in the following years. The clover seed is collected and sold to a seed company for cleaning and resale. Silage produced on the leys provides food for livestock over the winter months. Other parts of the rotation include wheat, oats, and peas (another legume), with fodder crops and naked barley among some of the other crops grown on the farm.

For additional fertilizer the farm takes in garden waste collected by the council from local towns to use as compost and this provides the largest external input to the farm. The composting process takes place on the farm and the resulting compost provides essential nutrients and organic matter to the soil.

It is not just the products that are diverse; the farm is also utilizing diversity within single crops. The farmer is one of a few UK farms trialling 'Population Wheat'—a heterogeneous wheat made from crosses of numerous varieties (in this case 'Wakelyns' population, which has 190 crosses of 20 varieties). Although in good

Figure 8.2 An agrecological approach, Shropshire, UK. Organic sheep graze on pea straw alongside newly planted walnut trees, before they are moved to pasture. Photo: Nicola Randall

Continued

> **Box 8.3** *(Continued)*
>
> conditions the mixture of different varieties may be less productive than a single strain bred solely for yield, the aim is to increase resilience to pests, diseases, and climatic extremes through increased genetic diversity. Undersowing and intercropping are practised on the farm to increase diversity further.
>
> This diverse and well-considered approach to management of the farm is extended to marketing, and many of the products are sold to local or specialist retailers, thus building resilience in marketing as well as the agroecosystem itself. Figure 8.2 shows a small section of the farm, where livestock, nut trees, arable crops, and pasture intersect.

8.5 Incentivizing a more agroecological approach

The implementation of conservation agriculture will depend heavily on the regulatory framework and associated incentives to support farmers making the transition. Transition from highly intensive farming to low-input sustainable management entails some financial risk and maintaining biodiversity long term can be a cost to the farmer, while many of the benefits are considered a public good. How this is delivered varies considerably around the world.

8.5.1 Funding

8.5.1.1 A European example

The Common Agricultural Policy (CAP) in Europe is a series of incentives aiming to support farmers and improve agricultural productivity, so that consumers have a stable supply of affordable food; ensure that European Union (EU) farmers can make a reasonable living; help tackle climate change and the sustainable management of natural resources; maintain rural areas and landscapes across the EU; and keep the rural economy alive by promoting jobs in farming, agri-foods industries, and associated sectors. The CAP delivers direct payments to farmers. In a recent reform of CAP, new 'greening' rules have been introduced which direct that 30 per cent of the Direct Payment envelope, paid per hectare, is linked to three environmentally friendly farming practices: crop diversification, maintaining permanent grassland, and conserving 5 per cent of areas of ecological interest or measures considered to have at least equivalent environmental benefit. How this is implemented is largely up to each member state. In England, the government has introduced a Countryside Stewardship Scheme (CSS) which aims to support farmers to conserve and restore wildlife habitats; manage flood risk; create and manage woodland; reduce widespread water pollution from agriculture; keep the character of the countryside; preserve historical features in the landscape; and encourage educational access. This scheme supports the implementation of conservation agriculture approaches as they deliver the aims of the CSS.

8.5.1.2 A USA example

The US federal government supports farmers providing environmental benefits on farmland with a range of grants, loans, and long-term rental agreements. These include a Conservation Stewardship Program, which rewards advanced conservation systems through 5-year renewable payment contracts to farmers and ranchers to actively manage, maintain, and expand conservation practices like rotations, cover cropping, rotational grazing, and IPM on working farms; a Conservation Reserve Program to protect soil, water quality, and habitat by removing highly erodible or environmentally sensitive land from agricultural production through long-term rental agreements; Conservation Loans, which offer federal guarantees on private loans to help farmers and ranchers implement conservation practices on their farm, with a priority for beginning farmers; and Conservation Innovation grants.

8.5.1.3 An Indian perspective

In low and middle income countries (LMICs) the situation is rather different. It varies hugely between countries and there is not space to describe all scenarios here. In India, aspects of conservation agriculture have been implemented for a long time and there have been efforts by extension agencies to promote organic approaches, minimum tillage, and IPM. However, farmers face many constraints, including lack of access to advice and the appropriate technology; not least, a supportive policy regime remains lacking (Raina et al. 2005; Bhan and Behera 2014).

The ability and willingness of governments to support conservation agriculture is key to its implementation and is an arena in which societal priorities have a clear link with the biology of agroecosystems and how they are managed.

8.6 Future perspectives

Conservation agriculture and sustainable approaches to farming are likely to be an important part of the future of farming. The Sustainable Development Goals (SDGs) prioritize Zero Hunger (SDG 2), Clean Water (SDG 6), Resposible Consumption and Production (SDG 12), Climate Action (SDG 13), and Life on Land (SDG 15). These all require that the negative impacts of intensive agricultural practice are reduced. Some of the reduction is likely to be delivered by novel technology, but equally, low-impact sustainable farming will play its part. An optimistic interpretation of this is that the consequence for the biology of the system is likely to be the recovery of biodiversity, delivered through a more holistic approach to agriculture supported by national governments. The knowledge of how is available; what is needed are accessible extension, resources, and policies to implement it.

9
Future Perspectives

9.1 Introduction

The future of food production will almost certainly be driven principally by a rising population, along with an increased demand for food in the context of climate change and resource constraints (OECD 2016). This is likely to put increasing pressures on agroecosystems and surrounding areas, which are already threatened by global drivers such as climate change, water availability, biodiversity declines, and associated ecosystem service losses.

A variety of methods have the potential to increase food outputs. For example, improving resource efficiency, increasing the agricultural land area, and expansion of irrigated areas can be combined with factors outside the agroecosystem, such as reducing food waste and post-harvest losses. Innovation in food and agriculture is a key element in meeting those challenges, but any increases in food production are likely to impact on the underlying biology of the agroecosystems.

9.2 Key drivers of agricultural change

9.2.1 Population growth

According to the United Nations 2017 revision of the 'World Population Prospects', the world's population in 2017 stood at 7.6 billion, with a projected rise to 8.6 billion in 2030, 9.8 billion in 2050, and 11.2 billion in 2100 (UN 2017). These estimates are based on a scenario where the rate of increase is declining (Figure 9.1).

The greatest increases in population are expected to take place in the least developed countries (LDC) as identified by the UN, with over half of this growth anticipated in Africa, followed by Asia, Latin America, and the Caribbean. In North America and Oceania, the population is expected to rise, but modestly. Only Europe is expected to have a falling population (UN 2017).

The Biology of Agroecosystems. Nicola P. Randall and Barbara Smith, Oxford University Press (2020).
© Nicola P. Randall and Barbara Smith 2020. DOI: 10.1093/oso/9780198737520.001.0001

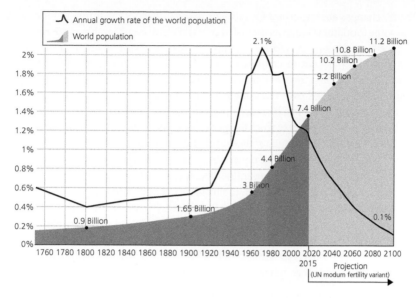

Figure 9.1 United Nations 2017 revision of the 'World Population Prospects'. https://ourworld indata.org/world-population-growth

In line with population trends, food demands are increasing globally. They are also shifting towards higher demands from the developing world (OECD 2016). Nine countries, led by India and Nigeria, have been identified as contributing the most to population increases. These high-growth countries will inevitably need to adjust their production systems, or significantly increase imports, leading to increased pressure on agroecosystems. Nigeria, for example, is largely an agrarian economy, with agriculture employing two-thirds of the workforce. However, in comparison to many other countries, productivity is relatively low, and increases in crop and livestock production have not kept pace with population increases. In 2017, less than 1 per cent of cropped land was irrigated, and around 60 per cent of ruminant livestock was pastoral and found in semi-arid regions. As a result, Nigeria imports a high proportion of its food. Policies are being put in place to improve productivity in Nigeria; alongside changes to marketing and extension, changes to production are required. Less reliance on rainfed agriculture, higher fertilizer applications, and different planting materials are likely to be introduced (FAO 2019a), all of which have the potential to impact on the diversity of wild species and natural ecosystems.

9.2.2 Changing consumer habits

Changing trends in consumption patterns are another factor that influences agroeco-systems. Consumer demands can be unpredictable. Demand for novel foods, which may be a passing fashion, can impact on farming systems in the countries that supply them, but some trends are more long term. Marketing, retailing, urbanization, and changes to incomes and to trade all influence trends in food consumption (Kearney 2010).

Dietary changes are important for estimating the impact of the future of food production; some food items (such as grain-fed meat) require considerably more resources to produce than others (per calorie). Cereals remain important in global diets (providing on average 50 per cent of calories in 2010), but meat consumption has seen huge increases—a threefold increase since the 1960s in developing countries—and egg consumption has doubled over the same period (Kearney 2010). Most studies anticipate that meat consumption will increase along with rising wealth and it has been predicted that there will be an increase from the per capita, per annum consumption of 32 kg in 2011 to 52 kg by 2050, though predicting changes to trends in diet and consumption is complex. There are high levels of uncertainly due to the interaction of economic drivers with cultural, social, and religious influences. It is uncertain whether developing economies such as Brazil and China will stabilize at levels of consumption similar to the UK and Europe or rise to those of the USA, which are higher. Irrespective of this, major increases in the consumption of meat, particularly grain-fed meat, are almost certain to impact the sustainability of food production through competition for land use, water, and other inputs (Fischer et al. 2013; Foresight 2011). This will in turn contribute to climate change through increased GHG emissions.

Although meat consumption is expected to rise, there are also calls for people to adopt plant-based diets; it is estimated their adoption could reduce global mortality by 6–10 per cent and food-related GHG emissions by 29–70 per cent (Springmann et al. 2016a). These calls are controversial, however, as livestock is an essential element of many mixed agricultural systems.

9.3 Future challenges and issues for the biology of agroecosystems

9.3.1 Competition for resources

The FAO considers competition for natural resources to be one of the key issues for agroecosystems, with increasing scarcity of land and water being key.

Agricultural expansion is still one of the main drivers of global deforestation, and agriculture is estimated to be responsible for around 80 per cent of deforestation. Although conversion of forests for agriculture has reduced since the turn of the century, around 7 million ha of tropical and subtropical forests were lost in the period between 2000 and 2010, alongside a 6 million ha expansion in agriculture. This leads to losses in biodiversity, soil degradation, and impacts on water quality. Competition between food production and biofuels also increases pressure on agroecosystems and on competition for land. Richer, higher yielding nations tend to increase crop production through intensification and yield increases, whereas poorer, lower yielding nations increase crop production through land clearing. If this trend trajectory continues, global land clearing will increase to approximately 1 million ha by 2050 (Tilman et al. 2011).

With agriculture accounting for 70 per cent of all water withdrawals, demand for water within agroecosystems also leads to competition for water. The FAO estimates that over 40 per cent of rural populations globally live in river basins classed as water scarce (where over 60 per cent of water is withdrawn). North America, Mediterranean Europe, South Africa, and South Asia suffer from water stress and it is estimated that this will expand into other regions in the future. North Africa and the Middle East are particularly vulnerable, with many countries withdrawing more than 60 per cent of their total fresh water for agriculture, industry, and cities, which are all competing for limited resources (FAO 2011. Furthermore, water quality is threatened by nutrient run-off and pollutants from poor waste water treatment (OECD 2016).

9.3.2 Climate change

Along with other drivers such as economic pressures combined with water availability, climatic factors impact on agricultural production. Changes to climate trends and climate variability have been well documented and are predicted by many scientists to continue. Climate change impacts agriculture, but agriculture also impacts climate change, with agriculture being attributed for up to one-third of anthropogenic GHG emissions (Vermeulen et al. 2012).

The projected impact of climate change on agricultural production varies with region, crop, and the level of climate change. In mid and high latitudes, production may increase and extend northwards under moderate climate change, and crops such as maize, sunflower, and soya that are currently grown in more southerly regions, such as southern Europe, may become viable in more northerly or higher regions. This contrasts with seasonally arid and tropical regions, where crops are already growing in the maximum viable temperatures, and even a 2°C temperature rise could reduce wheat production by 10 per cent in low latitudes (Gornall et al. 2010). Climate change impacts on production are likely to be caused by a variety of factors, the main ones being changes to temperature and precipitation.

Most current agricultural systems are rainfed, and as crops depend on water for growth, changes to precipitation and evaporation caused by temperature changes influence production. A change to the mean annual rainfall of only one standard deviation can be associated with a 10 per cent change in millet in South Asia, for example. Changes to season rainfall patterns may have an even greater effect on production than mean rainfall, through the influence of drought or waterlogging at key points in the growing cycle (Gornell et al. 2010).

Although drought is the most significant global environmental stress on agriculture (see Chapter 6), in many temperate regions predicted temperature rises are likely to be more of a challenge. In northern Europe, for example, water availability is not often an issue, but an increase in temperature above 27°C during flowering periods is likely to have a significant impact on wheat yields (Semenov and Shewry 2011).

Insect pest damage is likely to increase under warming. Insect metabolisms (and thus their food consumption rates) rise with increased temperatures. Population growth rates of insects also change with temperature, and although the direction of this change is likely to vary across the globe, temperate regions, where most cereal production takes place, are thought to be particularly vulnerable, as in these regions insect populations are more likely to rise (Deutsch et al. 2018). For example, the overwintering mortality of aphids is reduced under warmer temperatures, and both aphids and weevil larvae are reported to respond well to increased CO_2 levels (Gornall et al. 2010). In contrast, increases in atmospheric CO_2 may have positive impacts on crop yields, either by increasing photosynthesis (in C3 crops) or by increasing the water use efficiency (e.g. in C4 crops).

Despite all the uncertainty over the various predicted effects of climate change, most scientists are concerned that by 2050, climate change will contribute to an overall reduction in food production. Recent models predict a 3.2 per cent reduction in food availability. Of this, most of the reduction will be in fruit and vegetable availability (an overall 4.0 per cent reduction compared with 0.7 per cent reduction of red meat) (Springmann et al. 2016b). But staple crops such as wheat, rice, maize, and soybean, which between them provide two-thirds of human calorific intake, are also expected to experience overall reduced yields under temperature increases (Zhao et al. 2017). Springmann et al. (2016b) associate reduced food production with over 500,000 climate-related deaths worldwide. Most of these climate-related deaths are projected to occur in South and East Asia.

9.3.3 Biodiversity losses

All ecosystems, including agroecosystems, rely on interactions between organisms to provide and support natural services such as pollination, natural pest control, and soil fertility. It is the distinct traits of different organisms that facilitate the provision of these ecosystem services (see Chapter 3 and 4 for more details), but biodiversity declines have been shown to impact on these services and are likely to continue doing so. A 2019 review by Sánchez-Bayo and Wyckhuys found that more than 40 per cent of insect species are declining, and this is mainly attributed to agricultural intensification, including loss of off-crop habitats and heavy use of pesticides. These declines impact on broader biodiversity, as many higher taxa rely on insects directly or indirectly for food, but are predicted to have much further-reaching impacts, including on agricultural production itself. Disruptions to natural pest control, pollination, nutrient cycling, and decomposition services will all impact on production, and although most of the studies in this review were from western Europe or North America, tropical and developing countries are thought to face the same issues (Sánchez-Bayo and Wyckhuys 2019).

9.3.4 Antimicrobial resistance

Antimicrobial agents (AMAs) are frequently used within agricultural systems, often for veterinary use for farmed animals, but also, in some countries, as growth promoters. These are commonly administered either orally or via animal drinking water, but there are increasing concerns that their use can cause the spread of antimicrobial-resistant bacteria, and antimicrobial-resistant genes, within the environment. This is an issue not only for non-target animals on farms, but also more widely outside of the agroecosystems, including potentially for humans as many of the AMAs have similar structures to those used in human medicine.

These resistant genes and bacteria can be spread via livestock manure, and are widely present in soils and water, and various studies have detected them on fruit and vegetables. There are also concerns that plants may take-up AMAs, although scientists have found it difficult to prove direct relationships between a presence in the environment of these agents and resistant bacteria/genes. Investigations into the factors that influence the spread, persistence, and communicability of antimicrobial-resistant genes and bacteria are ongoing due to the potential risks to animal and human health (Thanner et al. 2016).

9.4 Potential solutions for sustainable intensification in agroecosystems

Many scientists and politicians agree that there is a need to preserve the areas of wilderness that remain (Watson et al. 2018). This, together with increased pressures of urbanization, means that the majority of increases in food production are likely to come from agricultural intensification and innovation rather than expansion. Although expansion still takes place, it is largely restricted to Africa and Latin America and the rate is slowing. Both economic and environmental constraints limit expansion into unproductive and marginal lands (OECD 2016). Sustainable intensification is the production of more food from the same or smaller areas of cultivated land with reduced inputs minimizing adverse environmental impacts. This includes complementary approaches such as integrating genetics with husbandry, precision farming, and ecological intensification.

9.4.1 Ecological approaches

9.4.1.1 Ecological intensification

Ecological Intensification (EI) is the intensification of ecological processes in the cultivated space to enhance performance (Caron et al. 2014). While not incompatible with high-tech approaches, EI focuses on managing the local environment to deliver ecosystem services in agriculture. The FAO defines it as 'a knowledge-intensive process that requires optimal management of nature's ecological functions and biodiversity to improve agricultural system performance, efficiency and farmers' livelihoods' (FAO 2019b) (http://www.fao.org/agriculture/crops/thematic-sitemap/theme/biodiversity/ecological-intensification/en).

EI is a young science. Central to the approach is the idea that biodiversity in farmed landscapes can provide benefits to production and crop health. Although relatively few studies are yet able to demonstrate the long-term impacts, there is emerging evidence across a number of countries that the approach can be effective. Gurr et al. (2016) showed that the introduction of nectar-producing plants around rice fields in Thailand, China, and Vietnam, over a period of 4 years, resulted in a lower abundance of two important pests, reduced insecticide applications by 70 per cent, increased grain yields by 5 per cent, and delivered an economic advantage of 7.5 per cent. They also recorded an increased abundance of beneficial insects, predators and parasitoids of rice pests, and detritivores when nectar-producing plants were present.

EI management tools include conservation tillage; integrated pest management; mixed cropping systems; diversified rotations; cover crops; direct seeding; improved nutrient management; conservation of biodiversity; and planting for synergistic effects (Wezel et al. 2015). Many of these interventions are already implemented under various agri-environment schemes and in conservation agriculture, but the key characteristic of EI is that it is designed specifically to target ecosystem services above and beyond biodiversity conservation and to support production (Kovács-Hostyánszki et al. 2017). Additionally, EI incorporates the notion that it is best implemented in conjunction with the integration of local knowledge into decision making, along with training and education of farmers in the effectiveness of the techniques, and a meaningful dialogue between farmers, scientists, and policy makers (Kovács-Hostyánszki et al. 2017).

9.4.1.2 Agroecology

Interest in agroecology is growing rapidly, yet its definition is contested. When it first emerged as a discipline in the 1980s, in response to the increasing intensification of agriculture through the green revolution, it was defined as '*the application of ecological concepts and principles to the design and management of sustainable agroecosystems, or the science of sustainable agriculture*' (Altieri 1995). Agroecology began as a farm-level approach, replacing inputs, using a similar model to that of organic farming, although chemical inputs were not completely banned, but instead a consideration of natural ecology was integrated with agriculture. However, through the 1990s it evolved to incorporate 'the ecology of the entire food system' (Francis et al. 2003) with an aim of bringing about full food system change, and included animal welfare considerations alongside natural ecology. In this context it has also taken a political economy focus (Gliessman 2018). In Latin America it has been developed as a movement and is associated with peasant and grass-roots community agriculture, respecting indigenous knowledge and local traditional techniques. For example, La Via Campesina is an international peasant rights movement which includes agroecology as the backbone of food sovereignty and environmental justice, challenging industrial agriculture and political structures. The approach is inherently low impact and favours diverse poly-cultures which support biodiversity in agroecosystems. In the European context the focus tends to remain on the practice rather than the politics (although not exclusively so), focusing on minimizing artificial inputs and habitat management; the impact of these two approaches, were they to be widely adopted, would have very different outcomes for the biology of the agroecosystem. One would result in a revolution in food production, with large numbers of highly engaged growers changing institutions and the political landscape around farming (with an uncertain outcome), while the other would result in a modification of current practice, more akin to EI.

9.4.1.3 Climate-smart agriculture

Climate-smart agriculture (CSA) is an approach that aims to sustainably increase agricultural productivity, while reducing GHG emissions and building resilience to climate change. This is largely a socio-economic approach that includes promoting evidence-building to support decisions, coordinated policy and organizational development, and improving financing to address food and climate security (Lipper et al. 2014). The details of how best to achieve this are still developing, but CSA has been accepted as a concept by global institutions such as the World Bank and FAO, and the integrated approach is likely to be supported by an agroecological approach, together with the development of new technologies.

9.4.2 Emerging innovations for agroecosystem biology

Modern innovations are not just designed to increase yield and farm incomes; many also aim to deal with global issues such as soil protection, climate change and other ecosystem losses, and food insecurity, but many remain controversial.

9.4.2.1 Elite rhizobia

Rhizobia are bacteria that form symbiotic nitrogen-fixing root nodules on legumes. Inoculating crops with rhizobia can increase yields and also contribute to soil nitrogen. Although principally associated with legumes, they have also found rhizobia in the endosphere and rhizosphere of cereals such as rice, and it has been shown that they can promote growth in non-leguminous plants (Checcucci et al. 2017). Inoculation with rhizobia is a common practice to aid plant establishment and growth. However, rhizobia are variable in their effectiveness. They are adapted to local conditions and some strains are more effective at promoting growth than others. In order to optimize their performance, rhizobial inoculants are screened for traits such as nitrogen fixation, edaphic adaptation, and performance in situ of rhizobial germplasm to identify elite strains with which to inoculate crops (Howieson et al. 2000). However, elite strains may be at a competitive disadvantage compared to indigenous strains, thereby reducing the effectiveness of inoculation under field conditions. To address this there is work in progress to improve strain selection and to genetically engineer superior nitrogen-fixing strains (Geetha and Joshi 2013).

9.4.2.2 Genetically modified crops

Modern technologies allow crops and other organisms to be genetically modified using biotechnology, and this can be used to deal with some of the issues within agricultural systems. Insecticidal, antibacterial, or antifungal properties can be used to reduce the amount of pesticide used in plants, for example. *Bacillus thuringiensis* (Bt) has been used in agriculture since the mid-1990s, but it can lead to yield penalties, as well as environmental issues such as the toxicity of transgenic pollen to non-target insects and the development of resistance in target insects. Bt crops are not the only way to enhance resistance of crops; various plastid transgenes have been shown to increase resistance to insect pests, but also to bacterial and viral pathogens. Genetic transformation can also be used to increase salt, drought, and temperature tolerance, although to date most experiments have been with tobacco plants rather than food plants. Another use for biotechnological engineering is improved productivity through better photosynthesis, or biofortification, manipulating a plant's nutrient content such as macro- or micronutrients. (Wani et al. 2015). Golden Rice, for example, is rice that has been genetically modified to biosynthesize beta-carotene, as an intervention for vitamin A deficiency, a serious public health issue in parts of Asia.

The first generation of genetically modified crops (generally referred to as GMOs) was created by transgenesis whereby foreign gene sequences from different non-sexually compatible plants (or other organisms) were inserted into the plant genome. In other words, the inserted sequences could not have occurred without artificial insertion. Examples include herbicide-tolerant soybean which incorporates a gene from the soil bacterium *Agrobacterium tumefaciens*; insect-resistant corn (specifically resisting the European corn borer) which incorporates the insecticidal protein Bt; resistance in plum to plum pox virus, which incorporates a protein from the virus itself; and vitamin-A-enriched rice, which incorporates genes for beta-carotene.

There has been much controversy surrounding the genetic modification of crops. In both the USA and the EU, GMOs must be labeled and there are regulatory frameworks in place around them. The pattern of uptake of GMOs varies globally. In Europe there has been more resistance than in the USA. In India, GMO Bt cotton (which resists cotton bollworm) now comprises 95 per cent of the cotton sown. The societal concerns around GMOs include the impact of the GMO on the local environment through the vertical transfer of GMO genes into the crop's wild-type counterparts; that the properties of the GMO can have an impact on non-target organisms and therefore biodiversity; and that newly developed organisms are effectively 'owned' by the companies that develop them, which introduces unpredictable economic and social consequences (Phillips 2008).

However, the technology is evolving and new engineered crops are being developed using CRISPR-Cas9 (usually abbreviated to CRISPR) which is a gene editing technique. The process allows for the modification of a single or a few nucleotides, targeting the insertion of genes to a specific location. The technique is powerful and can lead to disease resistance and improve metabolic function, offering great potential to increase yield. Regulation surrounding new generation GMOs is still in development. As foreign DNA is not inserted in the CRISPR process, these are not recognized as GMOs in the USA and are therefore currently unregulated. In the EU this is still being debated. The adoption of this technique in crop improvement and the future of next-generation GMOs will depend heavily on the policy frameworks that are developed around them. Technologically, there is huge potential, although the risks are largely unexplored.

9.4.2.3 Precision agriculture using new technologies

The aim of precision agriculture is to target management within fields to deliver inputs where they are needed, at the time they are needed, in order to optimize crop production, minimize the negative impact on the environment, and maximize economic returns. Using GPS and SatNav systems, farmers are able to map their farms to account for spatial variability in crop yield, topography, organic matter content, moisture levels, and various nutrient levels (McBrathney and Pringle 1999). Recognizing the natural variability, a farmer can then use smart technology to create a farm plan to customize treatments such as planting, watering, weeding, and pesticide application.

An emerging technology is the use of robotics for remote sensing and to replace artificial inputs. For example, machine vision can be used to recognize different species of weeds, to measure their biomass and measure the leaf area. It is then possible to either target small amounts of herbicide on weed leaves, thereby reducing chemical application by over 99 per cent, or to recognize the growing point of the weed and remove it with laser technology (Thomasson et al. 2017). As technologies develop, collaborative cooperation of drones with vehicles will be possible, with fully autonomous operations that may include all stages from inspections to spraying.

Similarly, proximal soil sensing (PSS) methods are being developed to allow farmers to determine variability in nutrients in soils across an individual field and target fertilizers optimally (Ji et al. 2017).

9.4.2.4 Nanotechnology in agriculture

Nanotechnology involves the use of particles with at least one dimension that is 100 nanometres (nm) or less. Nano-particles have a very large surface area relative to their mass, and this impacts on their properties. Nanotechnologies are receiving increasing recognition, and although at the time of writing are little used in agriculture, they have the potential to offer various uses within agroecosystems. Nano-encapsulated fertilizers and pesticides (insecticides, herbicides, and fungicides) are likely to see increased use in future as they may allow slow and sustained release of nutrients and agrochemicals and even allow smart delivery of active ingredients. They may also offer the potential to be used as diagnostic devices such as nanosensors that monitor plant health and growth or environmental conditions, or be used to support plant breeding and in genetic modifications (Parisi et al. 2015; Duhan et al. 2017).

9.4.3 Taking agriculture out of the fields: alternative food production systems

Many new technologies and innovations have involved removing food production from traditional agricultural systems. These may be integrating land- and water-based systems such as in aquaponics systems (see Chapter 6), integrating traditional food production with other types of production systems such as agroforestry (see Chapter 8), or taking food production into completely new environments. A selection of these is presented in the following sections.

9.4.3.1 Aeroponics

Aeroponics refers to the method of growing plants without soil and with very little water. Plants are grown vertically as well as horizontally, so making this process good for small spaces or urban agriculture, and it may even offer a viable food production method in space. NASA carried out experiments on the Mir Space Station and found that the Asian bean seedlings grown using aeroponic technology on the station in zero gravity grew better than the same plants on Earth.

One advantage of using aeroponic systems is that they typically use little to no growing media. The roots get maximum oxygen, and the plants grow more rapidly as a result. Aeroponic systems generally use less water than other types of hydroponic system (especially in true aeroponic systems in which a high-pressure, fine mist is provided to the plants). Harvesting is usually easier, especially for root crops as the roots can be easily accessed.

Aeroponics systems are costly to build and to maintain. There are also inherent technical challenges: for example, the mister/sprinkler heads can clog from build-up of the dissolved mineral elements in the nutrient solution. Additionally, because the plant's roots are hanging in mid-air in aeroponic systems, they are much more vulnerable to desiccation if there is any interruption in the watering cycle. Therefore even temporary power outage (for any reason) can lead to plant death, and this happens more quickly in aeroponic systems than in other types of hydroponics. Also there is a reduced margin for error with the nutrient levels in aeroponic systems, especially the true high-pressure systems.

9.4.3.2 Insect farming

Insect farming is a recent farming innovation in the West, although consumption (and to some extent production) of invertebrates is common in some countries. There is the potential for insect farming to provide advantages throughout food supply chains by the efficient creation of protein, less food waste, and, if well managed, less air and water pollution, fewer livestock welfare issues, lower risk of diseases, and reduced environmental consequences of intensive farming.

Western diets already include invertebrates; there are shrimp, lobsters, and crabs on supermarket shelves, and there is potential for other arthropods, including insects, to join them. In many regions, arthropods are already readily eaten and many are considered delicacies. Approximately 3,000 ethnic groups have been recorded as practising entomophagy (the practice of eating insects) across the world, particularly in Central and Latin America, Africa, Asia, Australia, and New Zealand (Ramos-Elorduy and Menzel 1998). The most commonly eaten orders of insects are beetles (Coleoptera; 31 per cent); butterflies and moths (Lepidoptera; 18 per cent); bees, wasps, and ants (Hymenoptera; 14 per cent); crickets and grasshoppers (Orthoptera; 13 per cent); true bugs (Hemiptera; 10 per cent); termites (Isoptera; 3 per cent); dragonflies (Odonata; 3 per cent); flies (Diptera; 2 per cent); and other orders (5 per cent). As examples, locusts, often fried, are eaten across the Middle East, in Africa, and parts of Asia; in Cambodia, spiders are popular and tarantulas are bred in holes, or collected from the forest, fried in oil and eaten as a nutritious snack; in Latin America, many species of ants are a popular food source (van Huis et al. 2013). Increasingly there is interest in the potential role of insects as a solution to food security, via direct consumption and via insects as a feed for other animals, as well as a potential source of oil to address fuel security. Box 9.1 explores different ways in which insects could be incorporated into food systems.

Box 9.1 Insects as food and commodities

The nutritional value of insects

Some studies report the crude protein content of insects is high (42–63 per cent), as is the lipid content (up to 36 per cent oil) (Makkar et al. 2014). However, there is some variation. A study in China investigated eleven insect orders and reported that they had higher protein contents than those in most plants and commercial meat, fowl, and eggs, ranging from 13 per cent to 77 per cent (Xiaoming et al. 2010). Other researchers found a range of 37–54 per cent protein content in eight insects found in Thailand (Raksakantong et al. 2010). For comparison, beef has, on average, a protein content of 26 per cent, chicken 27 per cent, fish 22 per cent, eggs 13 per cent, and milk 3.4 per cent. Insects are also high in terms of fats and oils, which are key to a healthy diet. Insect fat content has been shown to range from 9.12 per cent to 67.25 per cent (on a dry weight basis) and include essential fatty acids, such as linoleic (omega-6) and linolenic (omega-3) acids (Tao and Li 2018).

Insects can also provide a suite of minerals and vitamins. Nutrient analysis of wild black soldier fly (BSF) larvae in Kenya found 40 per cent crude protein, 33 per cent crude fat, 15 per cent ash, 12 per cent crude fibre, 0.56 per cent manganese, 3.07 per cent sodium, 0.57 per cent iron, 2.27 per cent potassium, and 0.24 mg of thiamin, 2.2 mg of riboflavin, and 1.3 mg of vitamin E per 100 g of weight (Nyakeri et al. 2017). Insects contain high levels of iron. Palm weevils and mopane caterpillars, both eaten in Africa, contain 12 and 31 mg of iron per 100 g of weight, respectively. In comparison, chicken and beef contain 1.2 and 3 mg of iron (Tao and Li 2018). As iron deficiency is one of the world's most common, widespread, and intractable micronutrient deficiencies, there is potential for a positive role in reducing 'hidden hunger' in developing countries.

Continued

Box 9.1 (*Continued*)

Use of insects in the food chain

Although insects are nutritious with high potential to contribute to food security, globally there are challenges around consumer acceptance in western cultures. It has been suggested that in western countries insect foods might be better accepted if introduced in processed foods, with women in particular being averse (Megido et al. 2016) . With time it is highly likely that this can be overcome.

An alternative to using insects in human diets is to use them within other parts of the food chain. Insect-based feed has been tested in pig, poultry, and fish production. Makkar et al. (2014) reviewed existing research and focused on five major insect species studied in relation to their products as animal feed: BSF larvae, house fly maggots, mealworm, locusts–grasshoppers–crickets, and silkworm meal. They found that maggot meal has been used for pig production in both Russia and Africa with no negative impact on either weight gain or breeding productivity. In rural Ghana, the addition of live maggots in poultry diets has been found to increase growth rate, clutch size, egg weight, eggs hatched, and chick weight, while in other experiments, including maggot meal in broiler chicken diet has shown potential, although a replacement for fish meal at 10 per cent is optimal as higher amounts appear to reduce palatability. In laying hens, partial replacement of the diet with maggot meal has been shown to have at best a positive impact on egg yield and hatchability and at worst no impact. This emphasis on partial, rather than complete, replacement is also reflected in research into the use of insect feed for fish, where both African catfish and tilapia have benefited from the inclusion of maggot meal, but total replacement reduces productivity. Nevertheless, there is clear potential for insects to play a role in livestock production, partially replacing soy, fish, and bone meal.

Insects as fuel

There is increasing concern over food–feed–fuel competition, and insect products could contribute to all three, including fuel. This is due to the high oil content of many insect species. For example, silk-worm pupae, a waste product from the silk industry that is often used as fertilizer or discarded, can be processed to produce oil that can then be used as a biofuel or in the production of other oil-based products. Many insect products are good at accumulating oils including BSF larvae, housefly maggot meal, and mealworm (Makkar et al. 2014).

Environmental and economic implications

Traditionally, insects are harvested wild from nature, and they are often used for home consumption or sold in local markets. An increase in agricultural intensification threatens invertebrate populations through land-use change and pesticide application, and this, coupled with new market demands, makes wild harvesting an impractical approach in the long term (Ramos-Elorduy 2006).

An alternative approach is to farm insects, which means producing them in an artificial environment either on small-scale farms or in large-scale industrialized rearing facilities. In order to assess the sustainability of insect farming, a number of factors must be considered (Figure 9.2).

Continued

Box 9.1 (*Continued*)

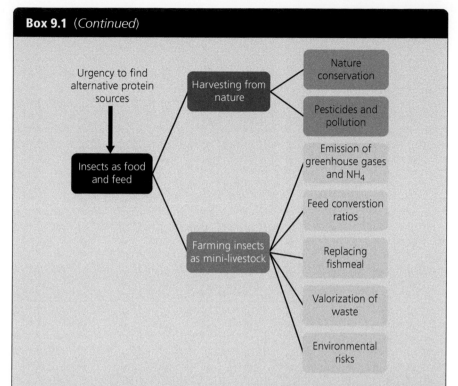

Figure 9.2 Environmental issues to be considered when insects are harvested or when reared as production animals or mini-livestock farms. From van Huis and Oonincx (2017)

A key metric of feed value is the feed-to-meat conversion efficiency (i.e. the quantity of feed that is needed to produce a 1-kg increase in weight). This has both environmental and economic implications, and varies according to the class of the animal and the production system, but in general, insects have a high feed conversion efficiency compared to mammals and birds due to their cold-blooded physiology. On average, insects can convert 2 kg of feed into 1 kg of insect mass, whereas cattle require 8 kg of feed to produce 1 kg of body weight gain (van Huis et al. 2013).

GHG emissions are another important consideration for sustainable agriculture. CO_2, CH_4, N_2O, and NH_3 (key GHGs) are emitted through the respiration and metabolism of insects and their feces. An analysis of five insect species shows that the emissions are lower than for pigs and beef cattle. However, this research is in its infancy and more detailed life cycle analyses (LCAs) looking at all aspects of production are needed. LCAs carried out for a limited number of insects indicate that insect rearing has a relatively low emission of GHGs (compared to livestock), but uses a high level of energy (as cold-bodied insects required heated rearing facilities).

As in conventional livestock systems, the principal source of environmental impact lies in the production of feedstock. In one example, the land use for mealworm

Continued

> **Box 9.1** (*Continued*)
>
> production was analyzed, and while the production facility accounted for 0.2 per cent of the total land use, the feed used in the facility was associated with 99 per cent of the land use (Oonincx and De Boer 2012). Similarly, the majority of water use is allocated to feed stocks, although again the water requirements are lower than in many traditional livestock systems. In a comparison with chicken, 1g of edible protein required 50 per cent more water and two to three times more land compared to mealworms and beef requires eight to fourteen times as much land and some five times as much water (Oonincx and de Boer 2012).
>
> Insect production may provide an opportunity to reuse organic waste, turning low-value waste into high-value protein. Some insects can be reared on substrates from waste streams including waste vegetable matter, waste from the brewing industry, rice straw, coffee pulp, and fish offal. The environmental impact of these novel approaches requires attention and there is room for innovation across the sector (van Huis and Oonincx 2017). The environmental impact of insect farming could be reduced further with novel designs for energy-efficient facilities, combined with efficient use of feed ingredients (van Huis and Oonincx 2017).

9.4.3.3 Meat substitutes

Production of meat substitutes is currently enjoying a surge in volume and has the potential to impact on the biology of agroecosystems through a reduction in land needed for livestock, land-use change, and potentially GHG emissions.

Meat substitutes provide certain characteristics such as the texture, flavour, and appearance of animal flesh. Traditionally available meat replacements include plants with meat-like qualities, or plants that have been processed, combined, or conventionally cooked (such as jackfruits, vegetable burgers, and tofu), but there are emerging meat substitutes that have been 'grown'. Food made from non-animal, non-plant, protein-rich masses grown in vats in an artificial environment with additions of chemicals/minerals include 'Quorn', which is made from textured mycoproteins derived from the *Fusarium venenatum* fungus and grown by fermentation. A more recent development is cultured meat grown in vitro from the muscle tissue of animals. Cyanobacteria are grown, sterilized, and fed, along with vitamins and growth factors, to in-vitro muscle cell cultures, and make up to 72 per cent of the weight of cultured cells. The stem cells are derived from animals but can be harvested by biopsy from them live, without significant invasion. A collagen matrix can be taken from live or dead animals. The growing culture requires a circulatory system to deliver oxygen and nutrients and to remove metabolic waste.

The idea of meat substitutes is not new: in 1932 Winston Churchill predicted that 'We shall escape the absurdity of growing a whole chicken in order to eat the breast or wing, by growing these parts separately under a suitable medium. Synthetic food will, of course, also be used in the future (Ford 2011). Cultured meat is not yet available to consumers but is feasible; in 2013 the first burger made from bovine stem cells was created.

There are energy conversion savings from direct consumption of plants rather than going through another trophic level, although conventional protein crops such as soy and lentil cannot be grown everywhere and so their transport costs must be accounted for. Quorn production (per kilo) has been estimated to use 140 times less land than beef production and produce 21 times less CO_2 (Nijdam et al. 2012). When compared to conventionally produced European meat, cultured meat involves approximately 7–45 per cent lower energy use (only poultry has lower energy use), 78–96 per cent lower GHG emissions, 99 per cent lower land use, and 82–96 per cent lower water use (depending on the product compared) (Tuomisto and Teixeira de Mattos 2011). Meat from in-vitro growth only takes weeks to reach harvest, compared to many months of farming of large animals such as pigs or cows. Cultured meat can be stored in the facility where it is grown, reducing the need for land, labour, and animal feed. Refrigeration and transport impacts (and associated carbon footprint) are lower too.

9.4.4 One answer is not enough

Each solution, when considered alone, is not enough to deal with the challenges and drivers facing future agroecosystems. Energy conversion savings from moving away from meat-based diets, for example, do not take into account the fact that the ruminant microbiome allows the conversion of indigestible plant material to protein-, vitamin-, and mineral-rich food sources for humans (Huws et al. 2018). New technologies have not yet enabled us to replicate this biome outside of the ruminant gut, nor have they yet enabled us to grow staple crops in high-saline areas, or to completely eradicate pest damage without impacting on non-target species. We are, however, taking small steps towards dealing with many of the conflicts within and around agriculture, and an understanding of the ecosystems themselves and the biology within them can help us make more informed choices regarding their management.

Glossary

Abiotic Non-living components of the ecosystem

Agrarian An agrarian society is one that is based around agriculture

Agriculture The science and practice of cultivating crops and rearing livestock to provide food and fuel

Agroecology A system of farming that makes use of ecological principles to enable sustainable food production

Agroecosystem The agroecosystem is a subset of global ecosystems (*see* Ecosystem). It is a unit area dominated by agricultural activity. The agroecosystem includes the region beyond the agricultural holding as this is impacted by agricultural activity. Typically, it is a matrix of farms and semi-natural areas

Allele A variant of a specific gene

Allelopathy An interference mechanism whereby one plant releases chemical compounds that negatively affect the growth of another at the same site. Known as allelochemicals, the compounds can be released in root exudates, decomposing tissues, or volatile compounds and can affect seed germination, growth, or survival of neighbouring plants

Arable Land used for growing crops

Biodiversity Biodiversity, or biological diversity, is a measure of the genetic and phenotypic variability among living organisms. The term biodiversity can be used to describe diversity at a variety of scales: within species, between species, and within ecosystems (as defined by the Convention on Biological Diversity)

Bio-geochemical cycle All the pathways by which a chemical can move through the biosphere, atmosphere, hydrosphere, and lithosphere

Biology The study of living things

Biotic Refers to living organisms. Biotic components are the living (or once living) components of a system and biotic processes are those that are mediated by living organisms

Biotope A region with uniform biological and environmental conditions supporting a characteristic community of plants and animals

Breed A group of individuals of one species that have been deliberately bred to express uniform characteristics

Complementarity Where niche differentiation (interspecific competition avoidance) allows a diversity of species to utilize more available resources overall in an environment

Crop A cultivated plant that is grown and harvested for food, fodder, or fuel

Crop rotation A management approach ensuring that the same crop is not sown in the same field in consecutive years, but that each crop is followed by another, preferably from a different plant family. Crop rotations are usually formalized in a sequence

Desertification The process whereby an area becomes increasingly arid. It can be a natural process or one driven by inappropriate agriculture

Domestication Adaptation of animals and plants over time, through selective breeding, to live in close association with, and for the benefit of, humans

Ecological intensification Application of ecological principles to maximize food production on an area of land sustainably

Ecosystem A community of living organisms, the abiotic environment, and the interactions that take place within it through flows of energy and nutrients within a given area

Ecosystem services The benefits ecosystems provide to humans

Eusocial A high level of social organization where roles are partitioned into reproduction and brood rearing, typically seen in insect colonies, with a single breeding female and non-reproductive workers

Evapotranspiration The sum of evaporation and plant transpiration from the Earth's land surface to the atmosphere

Evenness A measure of diversity: the relative proportion of species or functional groups at a given site

Evolution A change in the relative frequency of alternative alleles in a population

Exotic Non-native

Extensive agriculture An agricultural system that uses small inputs per hectare of labour, chemicals, and capital

Facilitation The process whereby the combination of organisms increases the resources required to the common pool

Food system An inclusive term that comprises the activities of food production, processing, and transport together with the governance and economics of the system

Functional redundancy Where more than one species fills the same ecological niche or performs the same function in a community or ecosystem

Functional traits Traits that influence ecosystem properties or species responses to environmental conditions

Genetic engineering Modification of the genetic material within an organism using biotechnology

GHG Greenhouse gases in the atmosphere which cause a rise in temperature by absorbing (and re-emitting earthwards) radiation that would otherwise be lost to space

Globalization The spread of products, processes, information, influence, interdependence, and functioning across national borders

GMO Genetically modified organism: where the genetic material of the organism has been changed by genetic engineering

Horticulture Literally 'garden-cultivation'; the cultivation of fruits, nuts, vegetables, and ornamental plants, excluding large-scale crop production or animal husbandry

Inorganic nutrients Nutrients that do not contain carbon in their molecular structure (largely minerals)

Invasive species A non-native introduced species that spreads and causes harm

IPM Integrated pest management: integration of a range of pest management techniques aimed at maintaining the level of pests below the threshold of economic injury, minimizing the use of artificial pesticides

Irrigation The application of controlled amounts of water to plants, typically by channels or pipes

Keystone species A species that impacts on multiple species or on an essential part of the system

Livestock Animals raised in an agricultural setting to obtain produce, e.g. meat, milk, and eggs

Mineralization The process of converting nutrients from organic to inorganic form

Monoculture The agricultural practice of growing/raising a single species (or variety) of crop/livestock over a large area of land

Mycorrhiza The symbiotic relationship between a fungus and a plant which takes place in the rhizosphere

Natural selection The process of adaptation whereby species with heritable traits that are well adapted to their environment survive and reproduce successfully to pass on those traits to their progeny. It is a key process in evolution

Nitrogen fixation The conversion of atmospheric nitrogen into a form that plants can use

Organic farming A method of farming that excludes inorganic fertilizers, pesticides, and GMOs. Organic farming relies on organic inputs such as manure and compost for nutrients and rotations and biological pest control

Organic nutrients Nutrients that contain carbon in their molecular structure, e.g. proteins, lipids, carbohydrates

Parasite An organism that develops on or within a host species. It may or may not kill the host, but usually reduces its fitness in some way

Parasitoid An organism that develops on or within a host during its immature lifestage. It always kills the host. The adult form is usually free living

Pastoral Primarily concerned with raising livestock

Pathogen A microorganism that causes disease

Perennial Living for several years. Perennial plants maintain at least the same rootstock in order to be considered the same plant

Phytohormone Plant hormone; hormones regulate cellular processes

Polyculture An agricultural system utilizing many crop types, designed to function more like a diverse natural ecosystem than a monoculture

Rainfed Rainfed agricultural systems are those that rely on precipitation for water

Resilience The ability of a system to absorb change and disturbance and still maintain the same relationships between populations or state variables

Rhizosphere The interface between plant roots and the soil

Richness A measure of diversity: the number of species or functional groups present in a given area

Riparian Alongside rivers

Rootstock Part of a plant, often underground, with established, healthy roots, from which new above-ground growth can be produced

Ruminants Animals (e.g. cows and sheep) that have a specialized feeding system containing multiple microbes, which enables them to break down plant material that is indigestible to humans

Salinization The process of accumulation of water-soluble salts in surface soils

Selective breeding The process of artificial selection, whereby humans identify desirable traits in animals and plants and selectively breed individuals with those traits to fix them in the population

Shifting cultivation A system of farming where small areas of land are carved out of naturally vegetated areas and cultivated for (usually) a single season, followed by several years of regeneration

Socioeconomic Relating to the interaction of social and economic factors

Species A group of individuals that can interbreed and produce viable offspring

Stress An organism's physiological response to a physical or psychological stressor. In plants this is defined as a biotic or abiotic constraint that limits the rate of photosynthesis and reduces the ability of the plant to convert energy into biomass (Grime 1977)

Sustainable The ability of the subject in question to continue in the same condition into the long-term future

Sustainable intensification The process whereby food production is maximized using only sustainable techniques. This may include techniques excluded from some agroecological and organic systems including GMOs and is not limited to implementing ecological principles. Typically it takes a systems approach

Tillage The preparation of the land for agriculture by mechanical means; it includes a range of techniques, from hand digging to industrial-scale ploughing

Transpiration The process of water movement within the plant; water is drawn into the plant via roots and leaves and water vapour is exhaled through leaf stomata

Trophic cascade Describes the indirect effect of a change in predator behaviour which releases prey from control and leads to an impact on organisms, communities, and processes further down the food chain

Varieties Subsets of species. In binomial classification, a third name indicates the variety

References

Abdulai, A. and CroleRees, A. 2001. Determinants of income diversification amongst rural households in southern Mali. Food Policy 26:437–52.

Aizen, M.A. and Harder, L.D. 2009. The global stock of domesticated honey bees is growing slower than agricultural demand for pollination. Current Biology 19:915–18.

Aktar, W., Sengupta, D., and Chowdhury, A. 2009. Impact of pesticides use in agriculture: their benefits and hazards. Interdisciplinary Toxicology 2:1–12.

Alexander, P., Rounsevell, M.D., Dislich, C., Dodson, J.R., Engström, K., and Moran, D. 2015. Drivers for global agricultural land use change: the nexus of diet, population, yield and bioenergy. Global Environmental Change 35:138–47.

Almaraz, M., Bai, E., Wang, C., Trousdell, J., Conley, S., Faloona, I., and Houlton, B.Z. 2018. Agriculture is a major source of NOx pollution in California. Science Advances 4:eaao3477.

Altieri, M.A. 1995. Agroecology—The Science of Sustainable Agriculture, 2nd edition. Westview Press, Boulder, CO.

Amezketa, E. 2006. An integrated methodology for assessing soil salinization, a precondition for land desertification. Journal of Arid Environments 67:594–606.

Asia and Pacific Commission on Agricultural Statistics. 2010. Report of the 23rd Session of the Asian and Pacific Commission on Agricultural Statistics. FAO, Rome.

Ausubel, J.H., Wernick, I.K., and Waggoner, P.E. 2013. Peak farmland and the prospect for land sparing. Population and Development Review 38(s1):221–42.

Avery, B. 1973. Soil classification in the Soil Survey of England and Wales. European Journal of Soil Science 24:324–38.

Backer, R., Rokem, S., Ilangumaran, G., Lamont, J., Praslickova, D., Ricci, E., Subramanian, S., and Smith, D.L. 2018. Plant growth-promoting rhizobacteria: context, mechanisms of action, and roadmap to commercialization of biostimulants for sustainable agriculture. Frontiers in Plant Science 9:1473.

Balmford, A., Green, R., and Phalan, B. 2012. What conservationists need to know about farming. Proceedings of the Royal Society B: Biological Sciences 279(1739):2714–24.

Barthlott, W., Hostert, A., Kier, G., Küper, W., Kreft, H., Mutke, J., Rafiqpoor, M.D., and Sommer, J.H. 2007. Geographic patterns of vascular plant diversity at continental to global scales (Geographische Muster der Gefäßpflanzenvielfalt im kontinentalen und globalen Maßstab). Erdkunde 61(4):305–15.

Barto, E. K., Weidenhamer, J.D., Cipollini, D., and Rillig, M.C. 2012. Fungal superhighways: do common mycorrhizal networks enhance below ground communication? Trends in Plant Science 17:633–7.

Bass, C., Denholm, I., Williamson, M.S., and Nauen, R. 2015. The global status of insect resistance to neonicotinoid insecticides. Pesticide Biochemistry and Physiology 121:78–87.

Basto, S., Thompson, K., Phoenix, G., Sloan, V., Leake, J., and Rees, M. 2015. Long-term nitrogen deposition depletes grassland seed banks. Nature Communications 6:6185.

Basu, P., Bhattacharya, R., and Ianetta, P. 2011. A decline in pollinator dependent vegetable crop productivity in India indicates pollination limitation and consequent agro-economic crises. Nature Preceedings http://precedings.nature.com/documents/6044/version/1.

Batary, P., Baldi, A., Sarospataki, M., Kohler, F., Verhulst, J., Knop, E., Herzog, F., and Kleijn, D. 2010. Effect of conservation management on bees and insect-pollinated grassland plant communities in three European countries. Agriculture, Ecosystems & Environment 136:35–9.

Baudron, F. and Giller, K.E. 2014. Agriculture and nature: trouble and strife? Biological Conservation 170:232–45.

Beauchemin, K.A., Kreuzer, M., O'Mara, F., and McAllister, T.A. 2008. Nutritional management for enteric methane abatement: a review. Australian Journal of Experimental Agriculture 48:21.

Bebber, D. P., Holmes, T. and Gurr, S.J. 2014. The global spread of crop pests and pathogens. Global Ecology and Biogeography 23:1398–407.

Bedoussac, L., Tribouillois, H., Plaza Bonilla, D., Journet, E.P., and Justes, E. 2017. Designing and evaluating arable cropping systems with cash and cover crop legumes in sole crop and intercrop to improve nitrogen use efficiency. Agricultural Research & Technology Open Access Journal 12(2):555842.

Bedoussac, L., Journet, E.P., Hauggaard-Nielsen, H., Naudin, C., Corre-Hellou, G., Jensen, E.S., and Justes, E. 2018. Grain legume–cereal intercropping systems. In: Achieving Sustainable Cultivation of Grain Legumes, eds S. Sivasankar, D. Bergvinson, P. Gaur, S.K. Agrawal, S. Beebe, and M. Tamò. Burleigh Dodds Science Publishing, Cambridge.

Beetz, A. and Rinehart, L. 2006. Pastures: Sustainable Management. National Centre for Appropriate Technology, Butte, MT.

Behnke, R. and Kerven, C. 2013. Counting the costs: replacing pastoralism with irrigated agriculture in the Awash Valley. In: Pastoralism and Development in Africa, eds A. Catley, J. Lind, and I. Scoones, pp. 82–95. Routledge, Abingdon.

Beman, J.M., Arrigo, K.R., and Matson, P.A. 2005. Agricultural runoff fuels large phytoplankton blooms in vulnerable areas of the ocean. Nature 434(7030):211.

Beneduzi, A., Ambrosini, A., and Passaglia, L.M. 2012. Plant growth-promoting rhizobacteria (PGPR): their potential as antagonists and biocontrol agents. Genetics and Molecular Biology 35(4):1044–51.

Bennett, E.M., Carpenter, S.R., and Caraco, N.F. 2001. Human impact on erodable phosphorus and eutrophication: a global perspective: increasing accumulation of phosphorus in soil threatens rivers, lakes, and coastal oceans with eutrophication. AIBS Bulletin 51:227–34.

Bennett, E.L., Eves, H.E., Robinson, J.G., and Wilkie, D.S. 2002. Why is eating bushmeat a biodiversity crisis? Conservation Biology in Practice 3:28–9.

Benton, T.G., Vickery, J.A., and Wilson, J.D. 2003. Farmland biodiversity: is habitat heterogeneity the key? Trends in Ecology & Evolution 18:182–8.

Bertone, M.A. 2004. Dung Beetles (Coleoptera: Scarabaeidae and Geotrupidae) of North Carolina Cattle Pastures and their Implications for Pasture Improvement. https://repository.lib.ncsu.edu/handle/1840.16/1952.

Bhan, S. and Behera, U.K. 2014. Conservation agriculture in India—problems, prospects and policy issues. International Soil and Water Conservation Research 2(4):1–12.

Bischoff, A. 2002. Dispersal and establishment of floodplain grassland species as limiting factors in restoration. Biological Conservation 104:25–33.

Blaauw, B.R. and Isaacs, R. 2014. Flower plantings increase wild bee abundance and the pollination services provided to a pollination-dependent crop. Journal of Applied Ecology 51:890–8.

Black, H. and Okwakol, M. 1997. Agricultural intensification, soil biodiversity and agroecosystem function in the tropics: the role of termites. Applied Soil Ecology 6:37–53.

Blitzer, E.J., Gibbs, J., Park, M.G., and Danforth, B.N. 2016. Pollination services for apple are dependent on diverse wild bee communities. Agriculture, Ecosystems & Environment 221:1–7.

Blouin, M., Hodson, M.E., Delgado, E.A., Baker, G., Brussaard, L., Butt, K.R., Dai, J., Dendooven, L., Pérès, G., Tondoh, J.E., and Cluzeau, D. 2013. A review of earthworm impact on soil function and ecosystem services. European Journal of Soil Science 64(2):161–82.

Blüthgen, N. and Feldhaar, H. 2010. Food and shelter: how resources influence ant ecology. In: Ant Ecology, eds L. Lach, C. Parr, and K. Abbott, pp. 115–36. Oxford Scholarship Online.

Boatman, N.D., Brickle, N.W., Hart, J.D., Milsom, T.P., Morris, A.J., Murray, A.W., Murray, K.A., and Robertson, P.A. 2004. Evidence for the indirect effects of pesticides on farmland birds. IBIS 146:131–43.

Bommarco, R., Marini, L., and Vaissière, B.E. 2012. Insect pollination enhances seed yield, quality, and market value in oilseed rape. Oecologia 169(4):1025–32.

Bommarco, R., Kleijn, D., and Potts, S.G. 2013. Ecological intensification: harnessing ecosystem services for food security. Trends in Ecology & Evolution 28(4):230–8.

Bonfoh, B., Schwabenbauer, K., Wallinga, D., Hartung, J., Schelling, E., Zinsstag, J., Meslin, F.X., Tschopp, R., Akakpo, J.A., and Tanner, M. 2010. Human health hazards associated with livestock production. In: Livestock in a Changing Landscape, Volume 1, Drivers, Consequences and Responses, eds H. Steinfeld, H. Mooney, and F. Schneider, pp. 197–220. Island Press, Washington, DC.

Borras Jr, S.M., Hall, R., Scoones, I., White, B., and Wolford, W. 2011. Towards a better understanding of global land grabbing: an editorial introduction. Journal of Peasant Studies 38:209–16.

Brereton, T.M., Botham, M.S., Middlebrook, I., Randle, Z., Noble, D., and Roy, D.B. 2017. United Kingdom Butterfly Monitoring Scheme Report for 2016. Centre for Ecology & Hydrology & Butterfly Conservation, Wallingford, UK.

Bronick, C.J. and Lal, R. 2005. Soil structure and management: a review. Geoderma 124:3–22.

Brooker, R.W., Bennett, A.E., Cong, W.F., Daniell, T.J., George, T.S., Hallett, P.D., Hawes, C., Iannetta, P.P., Jones, H.G., Karley, A.J., and Li, L. 2015. Improving intercropping: a synthesis of research in agronomy, plant physiology and ecology. New Phytologist 206(1):107–17.

Brown, L.R. 2006. The Earth Is Shrinking: Advancing Deserts and Rising Seas Squeezing Civilization. Earth Policy Institute, Washington, DC.

Brown, L.R. 2012. Outgrowing the Earth: The Food Security Challenge in an Age of Falling Water Tables and Rising Temperatures. Routledge, Abingdon.

Brown, P.H., Welch, R.M., and Cary, E.E. 1987. Nickel: a micronutrient essential for higher plants. Plant Physiology 85:801–3.

Bruinsma, J. 2017. World Agriculture: Towards 2015/2030: An FAO Study. Routledge, Abingdon.

Budge, G.E., Hodgetts, J., Jones, E.P., Ostojá-Starzewski, J.C., Hall, J., Tomkies, V., Semmence, N., Brown, M., Wakefield, M., and Stainton, K. 2017. The invasion, provenance and diversity of Vespa velutina Lepeletier (Hymenoptera: Vespidae) in Great Britain. PLoS One 12(9):e0185172.

Burgess, P. 2017. Agroforestry in the UK. Quarterly Journal of Forestry 111(2):111–16.

Burke, I. and Bell, J. 2014. Plant health management: herbicides. In: Encyclopedia of Agriculture and Food Systems, ed. N.K. van Alfen, pp. 425–40. Elsevier, Amsterdam.

Calvo, F.J., Knapp, M., van Houten, Y.M., Hoogerbrugge, H., and Belda, J.E. 2015. *Amblyseius swirskii*: what made this predatory mite such a successful biocontrol agent? Experimental and Applied Acarology 65:419–33.

Cameron, D., Osborne, C., Horton, P., and Sinclair, M. 2015. A sustainable model for intensive agriculture. Grantham Institute, York.

Campos, P., Ovando, P., Mesa, B., and Oviedo, J.L. 2018. Environmental income of livestock grazing on privately-owned silvopastoral farms in Andalusia, Spain. Land Degradation & Development 29(2):250–61.

Cao, L., Wang, W., Yang, Y., Yang, C., Yuan, Z., Xiong, S., and Diana, J. 2007. Environmental impact of aquaculture and countermeasures to aquaculture pollution in China. Environmental Science and Pollution Research International 14(7):452–62.

Cardinale, B.J., Matulich, K.L., Hooper, D.U., Byrnes, J.E., Duffy, E., Gamfeldt, L., Balvanera, P., O'Connor, M.I., and Gonzalez, A. 2011. The functional role of producer diversity in ecosystems. American Journal of Botany 98:572–92.

Caron, P., Biénabe, E., and Hainzelin, E. 2014. Making transition towards ecological intensification of agriculture a reality: the gaps in and the role of scientific knowledge. Current Opinion in Environmental Sustainability 8:44–52.

Catley, A., Admassu, B., Bekele, G., and Abebe, D. 2014. Livestock mortality in pastoralist herds in Ethiopia and implications for drought response. Disasters 38(3):500–16.

Chamberlain, D.E., Joys, A.C., Johnson, P.J., Norton, L.R., Feber, R.E., and Fuller, R.J. 2010. Does organic farming benefit farmland birds in winter? Biology Letters 6:82–4.

Chauhan, B.S. and Johnson, D.E. 2009. Influence of tillage systems on weed seedling emergence pattern in rainfed rice. Soil and Tillage Research 106(1):15–21.

Checcucci, A., DiCenzo, G.C., Bazzicalupo, M., and Mengoni, A. 2017. Trade, diplomacy, and warfare: the quest for elite rhizobia inoculant strains. Frontiers in Microbiology 8:2207.

Chiarelli, D.D., Rosa, L., Rulli, M.C., and D'Odorico, P. 2018. The water–land–food nexus of natural rubber production. Journal of Cleaner Production 172:1739–47.

Cleland, E.E. 2011. Biodiversity and ecosytem stability. Nature Education Knowledge 3(10):14.

Clements, D.R., Larsen, T., and Grenz, J. 2016. Knotweed management strategies in North America with the advent of widespread hybrid Bohemian knotweed, regional differences, and the potential for biocontrol via the psyllid *Aphalara itadori Shinji*. Invasive Plant Science and Management 9(1):60–70.

Collins, A.L., Anthony, S.G., Hawley, J., and Turner, T. 2015. Predicting potential change in agricultural sediment inputs to rivers across England and Wales by 2015. Marine and Freshwater Research 60:626–37.

Cong, W.F., Hoffland, E., Li, L., Six, J., Sun, J.H., X. G. Bao, F. S. Zhang, and Van Der Werf, W. 2015. Intercropping enhances soil carbon and nitrogen. Global Change Biology 21:1715–26.

Cordell, D., Drangert, J.-O., and White, S. 2009. The story of phosphorus: global food security and food for thought. Global Environmental Change 19:292–305.

Costanza, R., d'Arge, R., De Groot, R., Farber, S., Grasso, M., Hannon, B., Limburg, K., Naeem, S., O'Neill, R.V., Paruelo, J., and Raskin, R.G. 1997. The value of the world's ecosystem services and natural capital. Nature 387(6630):253.

Craswell, E.T., Grote, U., Henao, J., Vlek, P.L.G. 2004. Nutrient Flows in Agricultural Production and International Trade: Ecological and Policy Issues. ZEF Discussion Papers on Development Policy, No. 78. University of Bonn, Center for Development Research (ZEF), Bonn.

Cressey, D. 2017. The bitter battle over the world's most popular insecticides. Nature News 551:156.

Cresswell, C.J., Cunningham, H.M., Wilcox, A., and Randall, N.P. 2018. What specific plant traits support ecosystem services such as pollination, bio-control and water quality protection in temperate climates? A systematic map. Environmental Evidence 7(1):2.

D'Angelo, C. and Wiedenmann, J. 2014. Impacts of nutrient enrichment on coral reefs: new perspectives and implications for coastal management and reef survival. Current Opinion in Environmental Sustainability 7:82–93.

Daliakopoulos, I. N., Tsanis, I. K., Koutroulis, A., Kourgialas, N.N., Varouchakis, A.E., Karatzas, G.P., and Ritsema, C.J. 2016. The threat of soil salinity: a European scale review. Science of the Total Environment 573:727–39.

Darkoh, M.B.K. 2003. Regional perspectives on agriculture and biodiversity in the drylands of Africa. Journal of Arid Environments 54(2):261–79.

Darwin, C. 1868. The Variation of Animals and Plants under Domestication. John Murray, London.

Daryanto, S., Wang, L., and Jacinthe, P.-A. 2016. Global synthesis of drought effects on maize and wheat production. PLoS One v11(5).

Davies, G. 2012. Organic Vegetable Production: A Complete Guide. Crowood Press, Marlborough.

Davies, G., Turner, B., and Bond, B. 2008. Weed Management for Organic Farmers, Growers and Smallholders. Crowood Press, Marlborough.

Davis, M.A., Chew, M.K., Hobbs, R.J., Lugo, A.E., Ewel, J.J., Vermeij, G.J., Brown, J.H., Rosenzweig, M.L., Gardener, M.R., Carroll, S.P., and Thompson, K. 2011. Don't judge species on their origins. Nature 474(7350):153

Davis, S.C., Ming, R., LeBauer, D.S., and Long, S.P. 2015. Toward systems-level analysis of agricultural production from crassulacean acid metabolism (CAM): scaling from cell to commercial production. New Phytologist 208(1):66–72.

de Bruyn, L. and Conacher, A.J. 1990. The role of termites and ants in soil modification—a review. Soil Research 28:55–93.

de Oliveira, A.B., Alencar, N.L.M., and Gomes-Filho, E. 2013. Comparison between the water and salt stress effects on plant growth and development. In: Responses of Organisms to Water Stress, ed. S. Akinci. IntechOpen, London.

de Schutter, O. 2011. How not to think of land-grabbing: three critiques of large-scale investments in farmland. Journal of Peasant Studies 38:249–79.

de Vries, F.T., Thébault, E., Liiri, M., Birkhofer, K., Tsiafouli, M.A., Bjørnlund, L., Jørgensen, H.B., Brady, M.V., Christensen, S., and de Ruiter, P.C. 2013. Soil food web properties explain ecosystem services across European land use systems. Proceedings of the National Academy of Sciences 110:14296–301.

de Vries, W., Leip, A., Reinds, G., Kros, J., Lesschen, J.P., and Bouwman, A. 2011. Comparison of land nitrogen budgets for European agriculture by various modeling approaches. Environmental Pollution 159:3254–68.

Deb, D. 2017. Folk rice varieties, traditional knowledge and nutritional security in South Asia. In: Agroecology, Ecosystems, and Sustainability in the Tropics, pp. 118–26. Studera Press, Delhi.

Degens, B. 1997. Macro-aggregation of soils by biological bonding and binding mechanisms and the factors affecting these: a review. Soil Research 35:431–60.

Delaney, M., ArchMiller, A.A., Delaney, D.P., Wilson, A.E., and Sikora, E.J. 2018. Effectiveness of fungicide on soybean rust in the southeastern United States: a meta-analysis. Sustainability 10(6):1–15.

Deng, L., Zhang, Z., and Shangguan, Z. 2014. Long-term fencing effects on plant diversity and soil properties in China. Soil and Tillage Research 137:7–15.

Deutsch, C.A., Tewksbury, J.J., Tigchelaar, M., Battisti, D.S., Merrill, S.C., Huey, R.B., and Naylor, R.L. 2018. Increase in crop losses to insect pests in a warming climate. Science 361(6405):916–19.

Dietrich, J.P., Schmitz, C., Müller, C., Fader, M., Lotze-Campen, H., and Popp, H. 2012. Measuring agricultural land-use intensity—a global analysis using a model-assisted approach. Ecological Modelling 232:109–18.

Dixon, J., Gulliver, A., and Gibbon, D. 2001. Farming Systems and Poverty. Improving Farmers' Livelihoods in a Changing World. FAO, Rome, and World Bank, Washington, DC.

D'Odorico, P., Carr, J.A., Laio, F., Ridolfi, L., and Vandoni, S. 2014. Feeding humanity through global food trade. Earth's Future 2:458–69.

Doebley, J.F., Gaut, B.S., and Smith, B.D. 2006. The molecular genetics of crop domestication. Cell 127:1309–21.

Donald, P.F., Green, R.E., and Heath, M.F. 2001. Agricultural intensification and the collapse of Europe's farmland bird populations. Proceedings of the Royal Society of London B: Biological Sciences 268(1462):25–9.

Dorioz, J.M., Wang, D., Poulenard, J., and Trevisan, D. 2006. The effect of grass buffer strips on phosphorus dynamics—a critical review and synthesis as a basis for application in agricultural landscapes in France. Agriculture, Ecosystems & Environment 117:4–21.

Duffy, J. E., Godwin, C.M., and Cardinale, B.J. 2017. Biodiversity effects in the wild are common and as strong as key drivers of productivity. Nature 549:261.

Duhan, J.S., Kumar, R., Kumar, N., Kaur, P., Nehra, K., and Duhan, S. 2017. Nanotechnology: the new perspective in precision agriculture. Biotechnology Reports 15:11–23.

Edwards, A. and Withers, P. 2008. Transport and delivery of suspended solids, nitrogen and phosphorus from various sources to freshwaters in the UK. Journal of Hydrology 350:144–53.

Ehleringer, J.R. and Cerling, T.E. 2002. C3 and C4 photosynthesis. Encyclopedia of Global Environmental Change 2:186–90.

Eisler, M.C., Lee, M., Tarlton, J.F., Martin, G.B., Beddington, J., Dungait, J., Greathead, H., Liu, J., Mathew, S., and Miller, H. 2014. Agriculture: steps to sustainable livestock. Nature 507:32.

Erisman, J.W., Sutton, M.A., Galloway, J., Klimont, Z., and Winiwarter, W. 2008. How a century of ammonia synthesis changed the world. Nature Geoscience 1:636.

Esquinas-Alcázar, J. 2005. Protecting crop genetic diversity for food security: political, ethical and technical challenges. Nature Reviews Genetics 6(12):946.

EU Commission 2012. Report: The Implementation of the Soil Thematic Strategy and Ongoing Activities. COM 2012:46. http://ec.europa.eu/environment/soil/three_en.htm.

European Union. 2018. Land Cover and Land Use. https://ec.europa.eu/agriculture/sites/agriculture/files/statistics/facts-figures/land-cover-use.pdf.

FAO. 2004. What Is Agrobiodiversity? Building on Gender, Agrobiodiversity and Local Knowledge. FAO, Rome.

FAO. 2011. The State of the World's Land and Water Resources for Food and Agriculture (SOLAW)—Managing Systems at Risk. FAO, Rome, and Earthscan, London.

FAO. 2012. FAOSTAT Database. http://www.fao.org/faostat/en/#data.

FAO. 2013. FAO Statistical Yearbook. http://www.fao.org/docrep/018/i3107e/i3107e.pdf.

FAO. 2016. Meat and Meat Products. http://www.fao.org/fileadmin/templates/est/COMM_MARKETS_MONITORING/Meat/Documents/FO_Meat_June_2016.pdf.

FAO. 2017. FAO Advisory Note on Fall Armyworm (FAW) in Africa. FAO, Rome.

FAO. 2018. The Contribution of Insects to Food Security, Livelihoods and the Environment. http://www.fao.org/docrep/018/i3264e/i3264e00.pdf.

FAO. 2019a. FAO in Nigeria. Nigeria at a Glance. www.fao.org/nigeria/fao-in-nigeria/nigeria-at-a-glance/en/.

FAO. 2019b. Changing Paradigms of Agriculture. http://www.fao.org/agriculture/crops/thematic-sitemap/theme/biodiversity/ecological-intensification/en.

Feliciano, D., Ledo, A., Hillier, J., and Nayak, D.R. 2018. Which agroforestry options give the greatest soil and above ground carbon benefits in different world regions? Agriculture, Ecosystems & Environment 254:117–29.

Feltham, H., Park, K., Minderman, J., and Goulson, D. 2015. Experimental evidence that wildflower strips increase pollinator visits to crops. Ecology and Evolution 5(16):3523–30.

Feng, S., Shi, X., Reidsma, P., Ma, X., and Qu, F. 2012. Agricultural non-point source pollution in Taihu Lake Basin, China. In: Land Use Policies for Sustainable Development: Exploring Integrated Assessment Approaches, eds D. McNeill, I. Nesheim, and F. Brouwer, p. 69. Edward Elgar, Cheltenham.

Fernandes, M., Fernandes, F., Silva, D., Picanço, K., Jham, G., and Alves, F. 2015. Resistance of tomato accessions from the horticulture germplasm bank to red spider mites: exploring mite preference patterns and antixenosis mechanism. International Journal of Pest Management 61:284–91.

Fischer, G., Shah, M., and van Velthuizen, H. 2002. Climate Change and Agricultural Vulnerability. Special Report. International Institute for Applied Systems Analysis, Laxenburg.

Fischer, J., Abson, D.J., Butsic, V., Chappell, M.J., Ekroos, J., Hanspach, J., Kuemmerle, T., Smith, H.G., and von Wehrden, H. 2013. Land sparing versus land sharing: moving forward. Conservation Letters 7(3):149–57.

Fisher, T.W., Bellows, T.S., Caltagirone, L.E., Dahlsten, D.L., Huffaker, C.B., and Gordh, G. eds. 1999. Handbook of Biological Control: Principles and Applications of Biological Control. Elsevier, Amsterdam.

Fitzgibbon, C.D. 1997. Small mammals in farm woodlands: the effects of habitat, isolation and surrounding land-use patterns. Journal of Applied Ecology 34(2):530–9.

Fitzherbert, E.B., Struebig, M.J., Morel, A., Danielsen, F., Brühl, C.A., Donald, P.F., and Phalan, B. 2008. How will oil palm expansion affect biodiversity? Trends in Ecology & Evolution 23:538–45.

Flint, M.L. and Dreistadt, S.H. 1998. Natural Enemies Handbook: The Illustrated Guide to Biological Pest Control (Vol. 3386). University of California Press, Berkeley.

Foley, J.A., DeFries, R., Asner, G.P., Barford, C., Bonan, G., Carpenter, S.R., Chapin, F.S., Coe, M.T., Daily, G.C., and Gibbs, H.K. 2005. Global consequences of land use. Science 309:570–4.

Foley, J.A., Ramankutty, N., Brauman, K.A., Cassidy, E.S., Gerber, J.S., Johnston, M., Mueller, N.D., O'Connell, C., Ray, D.K., West, P.C. and Balzer, C., 2011. Solutions for a cultivated planet. Nature, 478(7369), p. 337.

Food and Drug Administration 2018. Antimicrobials Sold or Distributed for Use in Food-producing Animals. FDA, White Oak, MD

Ford, B.J. 2011. Impact of cultured meat on global agriculture. World Agriculture 2(2):43–6.

Forde, B.G. and Clarkson, D.T. 1999. Nitrate and ammonium nutrition of plants: physiological and molecular perspectives. Advances in Botanical Research 30:1–90.

Foresight. 2011. The Future of Food and Farming. Government Office for Science, London.

Forster, D., Andres, C., Verma, R., Zundel, C., Messmer, M.M., and Bäder, M. 2013. Yield and economic performance of organic and conventional cotton-based farming systems—results from a field trial in India. PLoS One 8(12):e81039.

Fox, J., Castella, J.C., Ziegler, A.D., and Westley, S.B. 2014. Expansion of rubber mono-cropping and its implications for the resilience of ecosystems in the face of climate change in Montane Mainland Southeast Asia. Global Environmental Research 18(2):145–50.

Francis, C., Lieblein, G., Gliessman, S., Breland, T.A., Creamer, N., Harwood, R., Salomonsson, L., Helenius, J., Rickerl, D., Salvador, R., and Wiedenhoeft, M. 2003. Agroecology: the ecology of food systems. Journal of Sustainable Agriculture 22(3):99–118.

Frison, E.A., Cherfas, J., and Hodgkin, T. 2011. Agricultural biodiversity is essential for a sustainable improvement in food and nutrition security. Sustainability 3:238–53.

Furumo, P.R. and Aide, T.M. 2017. Characterizing commercial oil palm expansion in Latin America: land use change and trade. Environmental Research Letters 12:024008.

Gaba, S., Bretagnolle, F., Rigaud, T., and Philippot, L. 2014. Managing biotic interactions for ecological intensification of agroecosystems. Frontiers in Ecology and Evolution 2:29.

Gabriel, D., Sait, S.M., Hodgson, J.A., Schmutz, U., Kunin, W.E., and Benton, T.G. 2010. Scale matters: the impact of organic farming on biodiversity at different spatial scales. Ecology Letters 13(7):858–69.

Gallai, N., Salles, J.M., Settele, J., and Vaissière, B.E. 2009. Economic valuation of the vulnerability of world agriculture confronted with pollinator decline. Ecological Economics 68(3):810–21.

Galloway, J.N. and Cowling, E.B. 2002. Reactive nitrogen and the world: 200 years of change. AMBIO 31(2):64–72.

Galloway, J.N., Dentener, F.J., Capone, D.G., Boyer, E.W., Howarth, R.W., Seitzinger, S.P., Asner, G.P., Cleveland, C.C., Green, P.A., Holland, E.A., and Karl, D.M. 2004. Nitrogen cycles: past, present, and future. Biogeochemistry 70:153–226.

Garibaldi, L.A., Steffan-Dewenter, I., Winfree, R., Aizen, M.A., Bommarco, R., Cunningham, S.A., Kremen, C., Carvalheiro, L.G., Harder, L.D., Afik, O., Bartomeus, I., Benjamin, F., Boreux, V., Cariveau, D., Chacoff, N.P., Dudenhoeffer, J.H., Freitas, B.M., Ghazoul, J., Greenleaf, S., Hipolito, J., Holzschuh, A., Howlett, B., Isaac, R., Javorek, S.K., Kennedy, C.M., Krewenka, K.M., Krishnan, S., Mandelik, Y., Mayfield, M.M., Motzke, I., Munyuli, T., Nault, B.A., Otieno, M., Petersen, J., Pisanty, G., Potts, S.G., Rader, R., Ricketts, T.H., Rundlof, M., Seymour, C.L., Schueepp, C., Szentgyoergyi, H., Taki, H., Tscharntke, T., Vergara, C.H., Viana, B.F., Wanger, T.C., Westphal, C., Williams, N., and Klein, A.M. 2013. Wild pollinators enhance fruit set of crops regardless of honey bee abundance. Science 339:1608–11.

Garratt, M.P.D., Truslove, C.L., Coston, D.J., Evans, R.L., Moss, E.D., Dodson, C., Jenner, N., Biesmeijer, J.C., and Potts, S.G. 2014. Pollination deficits in UK apple orchards. Journal of Pollination Ecology 12:9–14.

Geetha, S.J. and Joshi, S.J. 2013. Engineering rhizobial bioinoculants: a strategy to improve iron nutrition. Scientific World Journal, doi.org/10.1155/2013/315890.

Gerber, P.J., Steinfeld, H., Henderson, B., Mottet, A., Opio, C., Dijkman, J., Falcucci, A., and Tempio, G. 2013. Tackling Climate Change through Livestock—A Global Assessment of Emissions and Mitigation Opportunities. FAO, Rome.

Gibbs, H.K. and Salmon, J.M. 2015. Mapping the world's degraded lands. Applied Geography 57:12–21.

Gibbs, H.K., Ruesch, A.S., Achard, F., Clayton, M.K., Holmgren, P., Ramankutty, N., and Foley, J.A. 2010. Tropical forests were the primary sources of new agricultural land in the 1980s and 1990s. Proceedings of the National Academy of Sciences 107:16732–7.

Gillings, S., Newson, S.E., Noble, D.G., and Vickery, J.A. 2005. Winter availability of cereal stubbles attracts declining farmland birds and positively influences breeding population trends. Proceedings of the Royal Society of London B: Biological Sciences 272:733–9.

Głąb, L., Sowiński, J., Bough, R., and Dayan, F.E. 2017. Allelopathic potential of sorghum (*Sorghum bicolor* (L.) Moench) in weed control: a comprehensive review. Advances in Agronomy 145:43–95.

Gliessman, S. 2018. Defining agroecology. Agroecology and Sustainable Food Systems 42(6):599–600.

Godfray, H.C., Beddington, J.R., Crute, I.R., Haddad, L., Lawrence, D., Muir, J.F., Pretty, J., Robinson, S., Thomas, S.M. and Toulmin, C. 2010. Food security: the challenge of feeding 9 billion people. Science 327:812–18.

Gomez, J.D., Denef, K., Stewart, C.E., Zheng, J., and Cotrufo, M.F. 2014. Biochar addition rate influences soil microbial abundance and activity in temperate soils. European Journal of Soil Science 65:28–39.

Gornall, J., Betts, R., Burke, E., Clark, R., Camp, J., Willett, K., and Wiltshire, A. 2010. Implications of climate change for agricultural productivity in the early twenty-first century. Philosophical Transactions of the Royal Society B: Biological Sciences 365: 2973–89.

Goulding, K., Jarvis, S., and Whitmore, A. 2008. Optimizing nutrient management for farm systems. Philosophical Transactions of the Royal Society B: Biological Sciences, 363(1491):667–680.

Green, R.E., Cornell, S.J., Scharlemann, J.P.W., and Balmford, A. 2005. Farming and the fate of wild nature. Science 307:550–5.

Grigg, D. 1972. The Agricultural Systems of the World. Cambridge University Press, Cambridge.

Grime, J.P. 1977. Evidence for the existence of three primary strategies in plants and its relevance to ecological and evolutionary theory. American Naturalist 111(982):1169–94.

Grime, J.P. 1997. Biodiversity and ecosystem function: the debate deepens. Science 277 (5330):1260–1.

Grote, U., Craswell, E.T., and Vlek, P.L.G. 2008. Nutrient and virtual water flows in traded agricultural commodities. In: Land Use and Soil Resources, eds A.K. Braimoh and P.L.G. Vlek, pp. 121–43. Springer Science+Business Media, Berlin.

Gruber, K. 2017. The living library agrobioidversity. Nature 544:8–10.

Grządziel, J. 2017. Functional redundancy of soil microbiota—does more always mean better? Polish Journal of Soil Science 50:75.

Gulvik, M. 2007. Mites (Acari) as indicators of soil biodiversity and land use monitoring: a review. Polish Journal of Ecology 55:415.

Gurr, G.M., Lu, Z., Zheng, X., Xu, H., Zhu, P., Chen, G., Yao, X., Cheng, J., Zhu, Z., Catindig, J.L., and Villareal, S. 2016. Multi-country evidence that crop diversification promotes ecological intensification of agriculture. Nature Plants 2(3):16014.

Gustavsson, J., Cederberg, C., Sonesson, U., Van Otterdijk, R., and Meybeck, A. 2011. Global Food Losses and Food Waste. FAO, Rome.

Haaland, C., Naisbit, R.E., and Bersier, L.F. 2011. Sown wildflower strips for insect conservation: a review. Insect Conservation and Diversity 4:60–80.

Haddaway, N.R., Brown, C., Eales, J., Eggers, S., Josefsson, J., Kronvang, B., Randall, N.P., and Uusi-Kämppä, J. 2018. The multifunctional roles of vegetated strips around and within agricultural fields. Environmental Evidence 7(1):14.

Hajek, A. 2004. An Introduction to Biological Control. Cambridge University Press, Cambridge.

Hajek, A.E. and Eilenberg, J. 2018. Natural Enemies: An Introduction to Biological Control. Cambridge University Press, Cambridge.

Hammer, K. 1984. Das domestikationssyndrom. Kulturpflanze 32:11–34.

Hartemink, A.E. 2006. Soil erosion: perennial crop plantations. In: Encyclopedia of Soil Science, ed. R. Lal, pp. 1613–17. Taylor & Francis, Abingdon, UK.

Hassan, K., Pervin, M., Mondal, F., and Mala, M. 2016. Habitat management: a key option to enhance natural enemies of crop pest. Universal Journal of Plant Science 4(4):50–7.

Haukos, D.A., Johnson, L.A., Smith, L.M., and McMurry, S.T. 2016. Effectiveness of vegetation buffers surrounding playa wetlands at contaminant and sediment amelioration. Journal of Environmental Management 181:552–62.

Hawlena, D., Strickland, M.S., Bradford, M.A., and Schmitz, O.J. 2012. Fear of predation slows plant-litter decomposition. Science 336:1434–8.

Hawkins, H.J. 2017. A global assessment of Holistic Planned Grazing™ compared with season-long, continuous grazing: meta-analysis findings. African Journal of Range & Forage Science 34(2):65–75.

Heap, I. 2014. Global perspective of herbicide-resistant weeds. Pest Management Science 70:1306–15.

Hernandez-Aguilera, J.N., Conrad, J.M., Gómez, M.I., and Rodewald, A.D. 2019. The economics and ecology of shade-grown coffee: a model to incentivize shade and bird conservation. Ecological Economics 159:110–21.

Hirsch, P.R., Jhurreea, D., Williams, J.K., Murray, P.J., Scott, T., Misselbrook, T.H., T. Goulding, K.W., and Clark, I.M. 2016. Soil resilience and recovery: rapid community responses to management changes. Plant and Soil 412:283–97.

Hoekstra, A.Y. and Chapagain, A.K. 2007. The water footprints of Morocco and the Netherlands: global water use as a result of domestic consumption of agricultural commodities. Ecological Economics 64:43–151.

Hoffmann, I. 2011. Livestock biodiversity and sustainability. Livestock Science 139(1):69–79.

Holden, J., Haygarth, P.M., MacDonald, J., Jenkins, A., Sapiets, A., Orr, H.G., Dunn, N., Harris, B., Pearson, P.L., McGonigle, D., and Humble, A. 2015. Agriculture's Impacts on Water Quality. Global Food Security. NERC Open Research Archive, http://nora.nerc.ac.uk/id/eprint/510550/.

Huang, H.T. and Yang, P. 1987. The ancient cultured citrus ant. BioScience 39:665–71.

Holland, J.M. 2004. The environmental consequences of adopting conservation tillage in Europe: reviewing the evidence. Agriculture, Ecosystems & Environment 103(1):1–25.

Holland, J.M., Hutchison, M.A.S., Smith, B., and Aebischer, N.J. 2006. A review of invertebrates and seed-bearing plants as food for farmland birds in Europe. Annals of Applied Biology 148:49–71.

Holling, C.S. 1973. Resilience and stability of ecological systems. Annual Review of Ecology and Systematics 4:1–23.

Holmes, R.J. and Froud-Williams, R.J. 2005. Post-dispersal weed seed predation by avian and non-avian predators. Agriculture, Ecosystems & Environment 105:23–7.

Holzschuh, A., Steffan-Dewenter, I., Kleijn, D., and Tscharntke, T. 2007. Diversity of flower-visiting bees in cereal fields: effects of farming system, landscape composition and regional context. Journal of Applied Ecology 44:41–9.

Homeier, J., Baez, S., Hertel, D., and Leuschner, C. 2017. Editorial: tropical forest ecosystem responses to increasing nutrient availability. Frontiers in Earth Science 5:27. doi:10.3389/feart.2017.00027.

Hooper, D.U., Chapin, F.S., Ewel, J.J., Hector, A., Inchausti, P., Lavorel, S., Lawton, J.H., Lodge, D.M., Loreau, M., Naeem, S., and Schmid, B. 2005. Effects of biodiversity on ecosystem functioning: a consensus of current knowledge. Ecological Monographs 75:3–35.

House of Commons. 1919. Parliamentary Papers, House of Commons and Command, p. 588. HMSO, London.

Howieson, J.G., Malden, J., Yates, R.J., and O'Hara, G.W. 2000. Techniques for the selection and development of elite inoculant strains of *Rhizobium leguminosarum* in southern Australia. Symbiosis (Rehovot) 28(1):33–48.

Huang, N., Enkegaard, A., Osborne, L.S., Ramakers, P.M., Messelink, G.J., Pijnakker, J., and Murphy, G. 2011. The banker plant method in biological control. Critical Reviews in Plant Sciences 30(3):259–78.

Hudson, B.D. 1994. Soil organic matter and available water capacity. Journal of Soil and Water Conservation 49(2):189–94.

Huws, S.A., Creevey, C.J., Oyama, L.B., Mizrahi, I., Denman, S.E., Popova, M., Muñoz-Tamayo, R., Forano, E., Waters, S.M., Hess, M., and Tapio, I. 2018. Addressing global ruminant agricultural challenges through understanding the rumen microbiome: past, present, and future. Frontiers in Microbiology 9:2161.

Idrisi, Z. 2005. The Muslim Agricultural Revolution and its Influence on Europe. Foundation for Science, Technology and Civilization, Manchester.

IIASA/FAO. 2012. Global Agro-ecological Zones (GAEZ v3.0). IIASA, Laxenburg, and FAO, Rome.

Ikazaki, K., Shinjo, H., Tanaka, U., Tobita, S., Funakawa, S., and Kosaki, T. 2011. "Fallow Band System," a land management practice for controlling desertification and improving crop production in the Sahel, West Africa. 1. Effectiveness in desertification control and soil fertility improvement. Soil Science and Plant Nutrition 57:573–86.

Janick, J. 2000. Ancient Egyptian agriculture and the origins of horticulture. Acta Horticulturae 582:23–39.

Jansa, J., Wiemken, A., and Frossard, E. 2006. The effects of agricultural practices on arbuscular mycorrhizal fungi. Geological Society, London, Special Publications 266:89–115.

Jeffery, S., Abalos, D., Prodana, M., Bastos, A.C., Van Groenigen, J.W., Hungate, B.A. and Verheijen, F. 2017. Biochar boosts tropical but not temperate crop yields. Environmental Research Letters, 12 (5): p.053001.

Jenny, H. 1994. Factors of Soil Formation: A System of Quantitative Pedology. McGraw Hill, New York.

Ji, W., Adamchuk, V., Chen, S., Biswas, A., Leclerc, M., and Rossel, R.V. 2017. The Use of Proximal Soil Sensor Data Fusion and Digital Soil Mapping for Precision Agriculture. https://hal.archives-ouvertes.fr/hal-01601278.

Jones Jr, J.B. 2016. Hydroponics: A Practical Guide for the Soilless Grower. CRC Press, Boca Raton, FL.

Kassa, H., Dondeyne, S., Poesen, J., Frankl, A., and Nyssen, J. 2017. Transition from forest-based to cereal-based agricultural systems: a review of the drivers of land use change and degradation in Southwest Ethiopia. Land Degradation & Development 28(2):431–49.

Kay, D., Crowther, J., Fewtrell, L., Francis, C., Hopkins, M., Kay, C., McDonald, A., Stapleton, C., Watkins, J., and Wilkinson, J. 2008. Quantification and control of microbial pollution from agriculture: a new policy challenge? *Environmental Science & Policy* 11:171–84.

Kearney, J. 2010. Food consumption trends and drivers. Philosophical Transactions of the Royal Society B: Biological Sciences 365(1554):2793–807.

Khater, H.F., Ali, A.M., Abouelella, G.A., Marawan, M.A., Govindarajan, A., Murugan, K., Abbas, R.Z., Vaz, N.P., and Benelli, G. 2018. Toxicity and growth inhibition potential of vetiver, cinnamon, and lavender essential oils and their blends against larvae of the sheep blowfly, *Lucilia sericata*. International Journal of Dermatology 57:449–57.

Khoury, C.K., Bjorkman, A.D., Dempewolf, H., Ramirez-Villegas, J., Guarino, L., Jarvis, A., Rieseberg, L.H., and Struik, P.C. 2014. Increasing homogeneity in global food supplies and the implications for food security. Proceedings of the National Academy of Sciences 111:4001–6.

Kirkham, M.B. 2005. Field capacity, wilting point, available water, and the nonlimiting water range. In: Principles of Soil and Plant Water Relations, pp. 101–15. Academic Press, Cambridge, MA.

Klein, A.-M., Vaissière, B.E., Cane, J.H., Steffan-Dewenter, I., Cunningham, S.A., Kremen, C., and Tscharntke, T. 2007. Importance of pollinators in changing landscapes for world crops. Proceedings of the Royal Society B 274(1608):303–13.

Koehler, A. 2008. Water use in LCA: managing the planet's freshwater resources. International Journal of Life Cycle Assessment 13(6):451–5.

Koh, L.P. and Wilcove, D.S. 2008. Is oil palm agriculture really destroying tropical biodiversity? Conservation Letters 1(2):60–4.

Kottek, M., Grieser, J., Beck, C., Rudolf, B., and Rubel, F. 2006. World map of the Köppen–Geiger climate classification updated. Meteorologische Zeitschrift 15(3):259–63.

Kovács-Hostyánszki, A., Espíndola, A., Vanbergen, A.J., Settele, J., Kremen, C., and Dicks, L.V. 2017. Ecological intensification to mitigate impacts of conventional intensive land use on pollinators and pollination. Ecology Letters 20(5):673–89.

Krause, T. and Ness, B. 2017. Energizing agroforestry: *Ilex guayusa* as an additional commodity to diversify Amazonian agroforestry systems. International Journal of Biodiversity Science, Ecosystem Services & Management 13(1):191–203.

Kremen, C., Williams, N.M., and Thorp, R.W. 2002. Crop pollination from native bees at risk from agricultural intensification. Proceedings of the National Academy of Sciences 99:6812–16.

Kremen, C., Iles, A., and Bacon, C. 2012. Diversified farming systems: an agroecological, systems-based alternative to modern industrial agriculture. Ecology and Society 17(4):44.

Kristensen, T.N. and Sørensen, A.C. 2005. Inbreeding—lessons from animal breeding, evolutionary biology and conservation genetics. Animal Science 80:121–33.

Krupinsky, J.M., Bailey, K.L., McMullen, M.P., Gossen, B.D., and Turkington, T.K. 2002. Managing plant disease risk in diversified cropping systems. Agronomy Journal 94:198–209.

Kumar, A., Kaul, V., and Shankar, U. 2018. Response of Brinjal (*Solanum melongena* Guen.) varieties for the resistant reaction against Brinjal shoot and fruit borer (BSFB) and red spider mites (RSM). Journal of Entomology and Zoology Studies 6(2):1369–73.

Kumar, V., Sharma, A., Soni, J.K., and Pawar, N. 2017. Physiological response of C3, C4 and CAM plants in changeable climate. The Pharma Innovation 6(9, Part B):70.

Kurose, D., Furuya, N., Tsuchiya, K., Tsushima, S., and Evans, H.C. 2012. Endophytic fungi associated with *Fallopia japonica* (Polygonaceae) in Japan and their interactions with *Puccinia polygoni-amphibii* var. *tovariae*, a candidate for classical biological control. Fungal Biology 116(7):785–91.

Lal, B.B. 2003. Excavations at Kalibangan, the Early Harappans, 1960–1969. Archaeological Survey of India, New Delhi.

Landis, D.A. 2017. Designing agricultural landscapes for biodiversity-based ecosystem services. Basic and Applied Ecology 18:1–12.

Lanz, B., Dietz, S., and Swanson, T. 2018. The expansion of modern agriculture and global biodiversity decline: an integrated assessment. Ecological Economics 144:260–77.

Larson, G., Piperno, D.R., Allaby, R.G., Purugganan, M.D., Andersson, L., Arroyo-Kalin, M., Barton, L., Climer Vigueira, C., Denham, T., Dobney, K., Doust, A.N., Gepts, P., Gilbert, M.P.T., Gremillion, K.J., Lucas, L., Lukens, L., Marshall, F.B., Olsen, K.M., Pires, J.C., Richerson, P.J., Rubio de Casas, R., Sanjur, O.I., Thomas, M.G., and Fuller, D.Q.

2014. Current perspectives and the future of domestication studies. Proceedings of the National Academy of Sciences 111:6139–46.

Laurance, W.F., Sayer, J., and Cassman, K.G. 2014. Agricultural expansion and its impacts on tropical nature. Trends in Ecology & Evolution 29(2):107–16.

Lawlor, D.W., Lemaire, G., and Gastal, F. 2001. Nitrogen, plant growth and crop yield. In: Plant Nitrogen, pp. 343–67. Springer, Berlin.

Leal Filho, W., Mandel, M., Al-Amin, A.Q., Feher, A., and Chiappetta Jabbour, C.J. 2017. An assessment of the causes and consequences of agricultural land abandonment in Europe. International Journal of Sustainable Development & World Ecology 24:554–60.

Lehmann, J., Rillig, M.C., Thies, J., Masiello, C.A., Hockaday, W.C., and Crowley, D. 2011. Biochar effects on soil biota—a review. Soil Biology and Biochemistry 43:1812–36.

Lemaire, G., Franzluebbers, A., de Faccio Carvalho, P.C., and Dedieu, B. 2014. Integrated crop–livestock systems: strategies to achieve synergy between agricultural production and environmental quality. Agriculture, Ecosystems & Environment 190:4–8.

Lever, C. 2001. The Cane Toad: The History and Ecology of a Successful Colonist. Westbury Academic & Scientific Publishing, Otley, UK.

Lichtenberg, E.M., Kennedy, C.M., Kremen, C., Batáry, P., Berendse, F., Bommarco, R., Bosque-Pérez, N.A., Carvalheiro, L.G., Snyder, W.E., Williams, N.M., and Winfree, R. 2017. A global synthesis of the effects of diversified farming systems on arthropod diversity within fields and across agricultural landscapes. Global Change Biology 23(11):4946–57.

Lipper, L., Thornton, P., Campbell, B.M., Baedeker, T., Braimoh, A., Bwalya, M., Caron, P., Cattaneo, A., Garrity, D., Henry, K., and Hottle, R. 2014. Climate-smart agriculture for food security. Nature Climate Change 4(12):1068.

Lisonbee, L.D., Villalba, J.J., Provenza, F.D., and Hall, J.O. 2009. Tannins and self-medication: implications for sustainable parasite control in herbivores. Behavioural Processes 82:184–9.

Liu, H., Carvalhais, L.C., Crawford, M., Singh, E., Dennis, P.G., Pieterse, C.M.J., and Schenk, P.M. 2017. Inner plant values: diversity, colonization and benefits from endophytic bacteria. Frontiers in Microbiology 8:2552.

Liu, S., Chen, J., Gan, W., Schaefer, D., Gan, J., and Yang, X. 2015. Spider foraging strategy affects trophic cascades under natural and drought conditions. Scientific Reports 5:12396.

Lockwood, J.L., Hoopes, M.F., and Marchetti, M.P. 2013. Invasion Ecology. John Wiley & Sons, Chichester.

López-Serna, R., Ernst, F., and Wu, L. 2016. Analysis of cinnamaldehyde and diallyl disulfide as eco-pesticides in soils of different textures—a laboratory-scale mobility study. Journal of Soils and Sediments 16(2):566–80.

Lovett, B. and St. Leger, R.J. 2018. Genetically engineering better fungal biopesticides. Pest Management Science 74:781–9.

Lund, M., Carter, P., and Oplinger, E. 1993. Tillage and crop rotation affect corn, soybean, and winter wheat yields. Journal of Production Agriculture 6:207–13.

Machovina, B., Feeley, K.J., and Ripple, W.J. 2015. Biodiversity conservation: the key is reducing meat consumption. Science of the Total Environment 536:419–31.

Macnicol, R.D. and Beckett, P.H.T. 1985. Critical tissue concentrations of potentially toxic elements. Plant and Soil 85:107–29.

Magdoff, F. and Van Es, H. 2009. Building Soils for Better Crops: Sustainable Soil Management, 3rd edn. Sustainable Agriculture Network Handbook Series 10. Sustainable Agriculture Publications, Waldorf, MD.

Magurran, A. 2004. Measuring Biological Diversity. Blackwell Science, Oxford.

Makkar, H.P., Tran, G., Heuzé, V., and Ankers, P. 2014. State-of-the-art on use of insects as animal feed. Animal Feed Science and Technology 197:1–33.

Maikhuri, R.K., Semwal, R.L., Rao, K.S., Nautiyal, S., and Saxena, K.G. 1997. Eroding traditional crop diversity imperils the sustainability of agricultural systems in Central Himalaya. Current Science 73(9):777–82.

Malik, A.I., Colmer, T.D., Lambers, H., Setter, T.L., and Schortemeyer, M. 2002. Short-term waterlogging has long-term effects on the growth and physiology of wheat. New Phytologist 153(2):225–36.

Mann, C. 2011. How the Potato Changed the World. Smithsonian Magazine. https://www.smithsonianmag.com/history/how-the-potato-changed-the-world-108470605/.

Marks, E.A.N., Mattana, S., Alcañiz, J.M., and Domene, X. 2014. Biochars provoke diverse soil mesofauna reproductive responses in laboratory bioassays. European Journal of Soil Biology 60:104–11.

Marschner, H. 1995. Mineral Nutrition of Higher Plants, 2nd edn. Academic Press, London.

Martin-Guay, M.O., Paquette, A., Dupras, J., and Rivest, D. 2018. The new green revolution: sustainable intensification of agriculture by intercropping. Science of the Total Environment 615:767–72.

Masters, D., Edwards, N., Sillence, M., Avery, A., Revell, D., Friend, M., Sanford, P., Saul, G., Beverly, C., and Young, J. 2006. The role of livestock in the management of dryland salinity. Australian Journal of Experimental Agriculture 46:733–41.

Masters, D.G., Benes, S.E., and Norman, H.C. 2007. Biosaline agriculture for forage and livestock production. Agriculture, Ecosystems & Environment 119:234–48.

Mazoyer, M. and Roudart, L. 2007. A History of World Agriculture: From the Neolithic Age to the Current Crisis. Routledge, Abingdon, UK.

McLaughlin, A. and Mineau, P. 1995. The impact of agricultural practices on biodiversity. Agriculture, Ecosystems & Environment 55:201–12.

Megido, R.C., Gierts, C., Blecker, C., Brostaux, Y., Haubruge, É., Alabi, T., and Francis, F. 2016. Consumer acceptance of insect-based alternative meat products in western countries. Food Quality and Preference 52:237–43.

Mekonnen, M. M. and Hoekstra, A.Y. 2012. A global assessment of the water footprint of farm animal products. Ecosystems 15:401–15.

Melander, A.L. 1914. Can insects become resistant to sprays? Journal of Economic Entomology 7:167–73.

Mengel, K., Kirkby, E.A., Kosegarten, H., and Appel, T. 2001. Principles of Plant Nutrition. Kluwer Academic, Dordrecht.

Millennium Ecosystem Assessment. 2005. Ecosystems and Human Well-being: Synthesis Report. Millennium Ecosystem Assessment. http://www.millenniumassessment.org/en/Synthesis.aspx.

Ming, R., VanBuren, R., Wai, C.M., Tang, H., Schatz, M.C., Bowers, J.E., Lyons, E., Wang, M.L., Chen, J., Biggers, E., and Zhang, J. 2015. The pineapple genome and the evolution of CAM photosynthesis. Nature Genetics 47:1435.

Molina, G.A., Poggio, S.L., and Ghersa, C.M. 2014. Epigeal arthropod communities in intensively farmed landscapes: effects of land use mosaics, neighbourhood heterogeneity, and field position. Agriculture, Ecosystems & Environment 192:135–43.

Montanarella, L. 2015. Govern our soils. Nature 528:32–3.

Montgomery, D.R. 2007. Dirt: The Erosion of Civilizations. University of California Press, Berkeley.

Mora, C., Tittensor, D.P., Adl, S., Simpson, A.G., and Worm, B. 2011. How many species are there on Earth and in the ocean?. PLoS Biology 9(8):e1001127.

Morris, A.J., Holland, J.M., Smith, B., and Jones, N.E. 2004. Sustainable Arable Farming For an Improved Environment (SAFFIE): managing winter wheat sward structure for skylarks *Alauda arvensis*. Ibis 146:155–62.

Morris, N.L., Miller, P.C.H., Orson, J.H., and Froud-Williams, R.J. 2010. The adoption of non-inversion tillage systems in the United Kingdom and the agronomic impact on soil, crops and the environment—A review. Soil and Tillage Research 108:1–15.

Muschler, R.G. 2001. Shade improves coffee quality in a sub-optimal coffee-zone of Costa Rica. Agroforestry Systems 51(2):131–9.

Muscutt, A.D., Harris, G.L., Bailey, S.W., and Davies, D.B. 1993. Buffer zones to improve water quality: a review of their potential use in UK agriculture. Agriculture, Ecosystems & Environment 45:59–77.

Newbold, T., Hudson, L.N., Hill, S.L., Contu, S., Lysenko, I., Senior, R.A., Börger, L., Bennett, D.J., Choimes, A., Collen, B., and Day, J. 2015. Global effects of land use on local terrestrial biodiversity. Nature 520(7545):45.

Nichols, E., Spector, S., Louzada, J., Larsen, T., Amezquita, S., and Favila, M.E. 2008. Ecological functions and ecosystem services provided by Scarabaeinae dung beetles. Biological Conservation 141:1461–74.

Nijdam, D., Rood, T., and Westhoek, H. 2012. The price of protein: review of land use and carbon footprints from life cycle assessments of animal food products and their substitutes. Food Policy 37(6):760–70.

Noirot, C. and Darlington, J.P. 2000. Termite nests: architecture, regulation and defence. In: Termites: Evolution, Sociality, Symbioses, Ecology, eds Y. Abe, D.E. Bignell, and T. Higashi, pp. 121–39. Springer, Berlin.

Nolte, K., Chamberlain, W., and Giger, M. 2016. International Land Deals for Agriculture. Fresh insights from the Land Matrix: Analytical Report II. Bern Open Publishing, Bern.

Norton, L.R., Johnson, P.J., Joys, A.C., Stuart, R.C., Chamberlain, D.E., Feber, R.E., Firbank, L.G., Manley, W.J., Wolfe, M.S., Hart, B., Mathews, F., Macdonald, D.W., and Fuller, R.J. 2009. Consequences of organic and non-organic farming practices for field, farm and landscape complexity. Agriculture, Ecosystems and Environment 129(1–3):221–7.

Nyakeri, E., Ogola, H., Ayieko, M., and Amimo, F. 2017. An open system for farming black soldier fly larvae as a source of proteins for smallscale poultry and fish production. Journal of Insects as Food and Feed 3:51–6.

Odling-Smee, F.J., Odling-Smee, H., Laland, K.N., Feldman, M.W., and Feldman, F. 2003. Niche construction: the neglected process in evolution. Princeton University Press, Princeton.

OECD. 2016. Alternative Futures for Global Food and Agriculture. OECD, Paris.

OECD. 2017. OECD-FAO Agricultural Outlook 2017–2026. OECD, Paris.

Oldeman, L.R. 1992. Global extent of soil degradation. In: ISRIC Bi-Annual Report 1991/1992, ISRIC–World Soil Information, pp. 19–36. ISRIC, Wageningen.

Oldenburg, E., Kramer, S., Schrader, S., and Weinert, J. 2008. Impact of the earthworm *Lumbricus terrestris* on the degradation of *Fusarium*-infected and deoxynivalenol-contaminated wheat straw. Soil Biology and Biochemistry 40:3049–53.

Oliveira, R.S., Castro, P.M.L., Dodd, J.C., and Vosátka, M. 2005. Synergistic effect of *Glomus intraradices* and *Frankia* spp. on the growth and stress recovery of *Alnus glutinosa* in an alkaline anthropogenic sediment. Chemosphere 60:1462–70.

Ollerton, J., Winfree, R., and Tarrant, S., 2011. How many flowering plants are pollinated by animals? Oikos 120(3):21–326.

Ontl, T.A. and Schulte, L.A. 2012. Soil carbon storage. Nature Education Knowledge 3(10):35.

Oonincx, D.G. and de Boer, I.J. 2012. Environmental impact of the production of meal-worms as a protein source for humans—a life cycle assessment. PLoS One 7(12):e51145.

Opiyo, F., Wasonga, O., Nyangito, M., Schilling, J., and Munang, R. 2015. Drought adaptation and coping strategies among the Turkana pastoralists of northern Kenya. International Journal of Disaster Risk Science 6(3):295–309.

Orford, K.A., Murray, P.J., Vaughan, I.P., and Memmott, J. 2016. Modest enhancements to conventional grassland diversity improve the provision of pollination services. Journal of Applied Ecology 53(3):906–15.

Oswald, A. and Ransom, J.K. 2001. *Striga* control and improved farm productivity using crop rotation. Crop Protection 20(2):113–20.

Otero, I., Marull, J., Tello, E., Diana, G.L., Pons, M., Coll, F., and Boada, M. 2015. Land abandonment, landscape, and biodiversity: questioning the restorative character of the forest transition in the Mediterranean. Ecology and Society 20(2):7.

Overton, M. 1996. Agricultural Revolution in England: The Transformation of the Agrarian Economy 1500–1850 (Vol. 23). Cambridge University Press, Cambridge.

Oxford English Dictionary 2004. Oxford English Dictionary online. Mount Royal University Library, Calgary.

Paini, D.R., Sheppard, A.W., Cook, D.C., De Barro, P.J., Worner, S.P., and Thomas, M.B. 2016. Global threat to agriculture from invasive species. Proceedings of the National Academy of Sciences 113:7575–9.

Parisi, C., Vigani, M., and Rodríguez-Cerezo, E. 2015. Agricultural nanotechnologies: what are the current possibilities? Nano Today 10(2):124–7.

Partap, U. and Ya, T. 2012. The human pollinators of fruit crops in Maoxian County, Sichuan, China: a case study of the failure of pollination services and farmers' adaptation strategies. Mountain Research and Development 32(2):176–86.

Paull, J. 2016. Organics olympiad 2016: global indices of leadership in organic agriculture. Journal of Social and Development Sciences 7(2):79–87.

Pawar, P., Sharma, R., Sharma, A.K., and Kumar, R. 2016. Biopesticides and environment. In: Modern Approaches to Environmental Biotechnology, eds R. Kumar and A.K. Sharma, pp. 1–258. Blackwell, Oxford.

Peigné, J., Vian, J.-F., Payet, V., and Saby, N.P. 2018. Soil fertility after 10 years of conservation tillage in organic farming. Soil and Tillage Research 175:194–204.

Pelosi, C. and Römbke, J. 2017. Enchytraeids as bioindicators of land use and management. Applied Soil Ecology 123:775–9.

Petit, S., Boursault, A., and Bohan, D.A. 2014. Weed seed choice by carabid beetle (Coleoptera: T, Carabidae): linking field measurements with laboratory diet assessments. European Journal of Entomology 111:615–20.

Pfiffner, L. and Wyss, E. 2004. Use of sown wildflower strips to enhance natural enemies of agricultural pests. In: Ecological Engineering for Pest Management: Advances in Habitat Manipulation for Arthropods, eds G. Gurr, S. Wratten, and M. Altieri, pp. 165–86. CABI, Wallingford.

Phillips, T. 2008. Genetically modified organisms (GMOs): transgenic crops and recombinant DNA technology. Nature Education 1(1):213.

Pilon-Smits, E.A., Quinn, C.F., Tapken, W., Malagoli, M., and Schiavon, M. 2009. Physiological functions of beneficial elements. Current Opinion in Plant Biology 12:267–74.

Pimentel, D. 2006. Soil erosion: a food and environmental threat. Environment, Development and Sustainability 8:119–37.

Pineda, A., Kaplan, I., and Bezemer, T.M. 2017. Steering soil microbiomes to suppress aboveground insect pests. Trends in Plant Science 22:770–8.

Pinto-Tomás, A.A., Anderson, M.A., Suen, G., Stevenson, D.M., Chu, F.S., Cleland, W.W., Weimer, P.J., and Currie, C.R. 2009. Symbiotic nitrogen fixation in the fungus gardens of leaf-cutter ants. Science 326:1120–3.

Pitcairn, C.E.R., Leith, I.D., Sheppard, L.J., Sutton, M.A., Fowler, D., Munro, R.C., and Wilson, S.T.D. 1998. The relationship between nitrogen deposition, species composition and foliar nitrogen concentrations in woodland flora in the vicinity of livestock farms. Environmental Pollution 102(S1):41–48.

Plantureux, S., Peeters, A., and McCracken, D. 2005. Biodiversity in intensive grasslands: effect of management, improvement and challenges. Agronomy Research 3:153–64.

Portmann, F.T., Siebert, S., and Döll, P. 2010. MIRCA2000—global monthly irrigated and rainfed crop areas around the year 2000: a new high-resolution data set for agricultural and hydrological modeling, Global Biogeochemical Cycles 24, GB1011, doi:10.1029/2008GB003435.

Post, W.M., Nichols, J.A., Wang, D., West, T.O., Bandaru, V., and Izaurralde, R.C. 2013. Marginal lands: concept, assessment and management. Journal of Agricultural Science 5(5):129.

Potts, S.G., Biesmeijer, J.C., Kremen, C., Neumann, P., Schweiger, O., and Kunin, W.E. 2010. Global pollinator declines: trends, impacts and drivers. Trends in Ecology & Evolution 25:345–53.

Power, A.G. 2010. Ecosystem services and agriculture: tradeoffs and synergies. Philosophical Transactions of the Royal Society B: Biological Sciences 365(1554):2959–71.

Primavera, J.H. 2006. Overcoming the impacts of aquaculture on the coastal zone. Ocean & Coastal Management 49(9–10):531–45.

Puri, A., Padda, K.P. and Chanway, C.P., 2017. Nitrogen-fixation by endophytic bacteria in agricultural crops: recent advances. In Nitrogen in Agriculture-Updates. IntechOpen.

Qian, Y., Song, K., Hu, T., and Ying, T. 2018. Environmental status of livestock and poultry sectors in China under current transformation stage. Science of the Total Environment 622–623:702–9.

Qin, B., Zhu, G., Gao, G., Zhang, Y., Li, W., Paerl, H.W., and Carmichael, W.W. 2010. A drinking water crisis in Lake Taihu, China: linkage to climatic variability and lake management. Environmental Management 45(1):105–12.

Qui, J. 2009. Where the rubber meets the garden. Nature 457:246–7.

Raina, R.S., Sulaiman, R., Hall, A.J., and Sangar, S. 2005. Policy and institutional requirements for transition to conservation agriculture: an innovation systems perspective. In: Conservation Agriculture: Status and Prospects, eds I.P. Abrol, R.K. Gupta, and R.K. Mallik, pp. 224–32. Centre for Advancement of Sustainable Agriculture, New Delhi.

Rajeswaran, J., Duraimurugan, P., and Shanmugam, P. 2005. Role of spiders in agriculture and horticulture ecosystem. Journal of Food Agriculture and Environment 3:147.

Raksakantong, P., Meeso, N., Kubola, J., and Siriamornpun, S. 2010. Fatty acids and proximate composition of eight Thai edible terricolous insects. Food Research International 43:350–5.

Ramachandran Nair, P., Mohan Kumar, B., and Nair, V.D. 2009. Agroforestry as a strategy for carbon sequestration. Journal of Plant Nutrition and Soil Science 172:10–23.

Raman, R. 2017. The impact of genetically modified (GM) crops in modern agriculture: a review. GM Crops & Food 8(4):195–208.

Ramirez, K.S., Geisen, S., Morriën, E., Snoek, B.L., and van der Putten, W.H. 2018. Network analyses can advance above-belowground ecology. Trends in Plant Science 23(9):759–68.

Ramos-Elorduy, J. 2006. Threatened edible insects in Hidalgo, Mexico and some measures to preserve them. Journal of Ethnobiology and Ethnomedicine 2:51.

Ramos-Elorduy, J. and Menzel, P. 1998. Creepy Crawly Cuisine: The Gourmet Guide to Edible Insects. Inner Traditions/Bear & Co. Park Street Press, Rochester, VT.

Rashid, M.I., Mujawar, L.H., Shahzad, T., Almeelbi, T., Ismail, I.M., and Oves, M. 2016. Bacteria and fungi can contribute to nutrients bioavailability and aggregate formation in degraded soils. Microbiology Research 183:26–41.

Ray, D.K., Mueller, N.D., West, P.C., and Foley, J.A. 2013. Yield trends are insufficient to double global crop production by 2050. PLoS One 8(6):e66428.

Rayns, F. and Rosenfeld, A. 2010. Green Manures—Effects on Soil Nutrient Management and Soil Physical and Biological Properties. Horticultural Development Council, Kenilworth, UK.

Read, J., Fletcher, T.D., Wevill, T., and Deletic, A. 2010. Plant traits that enhance pollutant removal from stormwater in biofiltration systems. International Journal of Phytoremediation 12(1):34–53.

Reddy, D.E. 2007. Impact of globalization on small farmers worldwide: implications on information transfer. Proceedings of World Library and Information Congress, 73rd IFLA General Conference and Council, 19–23 August 2007, Durban, pp. 19–23.

Reganold, J.P. and Wachter, J.M. 2016. Organic agriculture in the twenty-first century. Nature Plants 2(2):15221.

Reichenberger, S., Bach, M., Skitschak, A., and Frede, H.G. 2007. Mitigation strategies to reduce pesticide inputs into ground- and surface water and their effectiveness: a review. Science of the Total Environment 384:1–35.

Reinhart, K.O., Packer, A., Van der Putten, W.H. and Clay, K. 2003. Plant–soil biota interactions and spatial distribution of black cherry in its native and invasive ranges. Ecology Letters, 6:1046-1050.

Richner, N., Holderegger, R., Linder, H.P., and Walter, T. 2014. Reviewing change in the arable flora of Europe: a meta-analysis. Weed Research 55(1):1–13.

Ridgeway, R.L., King, E.G., and Carrillo, J.L. 1977. Augmentation of natural enemies for control of plant pests in the western hemisphere. In: Biological Control by Augmentation of Natural Enemies, eds R.L. Ridgway and S.B. Vinson, pp. 379–416. Springer, Boston, MA.

Riley, H., Pommeresche, R., Eltun, R., Hansen, S. and Korsaeth, A. 2008. Soil structure, organic matter and earthworm activity in a comparison of cropping systems with contrasting tillage, rotations, fertilizer levels and manure use. Agriculture, Ecosystems & Environment 124:275–84.

Ripple, W.J., Wolf, C., Newsome, T.M., Galetti, M., Alamgir, M., Crist, E., Mahmoud, M.I., and Laurance, W.F. 2017. World scientists' warning to humanity: a second notice. BioScience 67:1026–8.

Robinson, T.P., Thornton, P.K., Franceschini, G., Kruska, R.L., Chiozza, F., Notenbaert, A., Cecchi, G., Herrero, M., Epprecht, M., Fritz, S., You, L., Conchedda, G., and See, L. 2011. Global Livestock Production Systems. FAO, Rome, and International Livestock Research Institute (ILRI), Nairobi.

Robinson, T.P., Wint, G.W., Conchedda, G., Van Boeckel, T.P., Ercoli, V., Palamara, E., Cinardi, G., D'Aietti, L., Hay, S.I., and Gilbert, M. 2014. Mapping the global distribution of livestock. PLoS One 9:e96084.

Rook, A.J. and Tallowin, J.R. 2003. Grazing and pasture management for biodiversity benefit. Animal Research 52:181–9.

Rosegrant, M.W., Ringler, C., and Zhu, T. 2009. Water for agriculture: maintaining food security under growing scarcity. Annual Review of Environment and Resources 34:205–22.

Roser, M. and Ritchie, H. 2017. Yields and Land Use in Agriculture. Our World in Data. https://ourworldindata.org/yields-and-land-use-in-agriculture#citation.

Roubik, D.W. ed. 1995. Pollination of Cultivated Plants in the Tropics (No. 118). FAO, Rome.

Rouland-Lefèvre, C. 2010. Termites as pests of agriculture. In: Biology of Termites: A Modern Synthesis, eds D.E. Bignell, Y. Roisin, and N. Lo, pp. 499–517. Springer, Dordrecht.

Rundlöf, M. and Smith, H.G. 2006. The effect of organic farming on butterfly diversity depends on landscape context. Journal of Applied Ecology 43:1121–7.

Rusch, A., Chaplin-Kramer, R., Gardiner, M.M., Hawro, V., Holland, J., Landis, D., Thies, C., Tscharntke, T., Weisser, W.W., Winqvist, C., and Woltz, M. 2016. Agricultural landscape simplification reduces natural pest control: a quantitative synthesis. Agriculture, Ecosystems & Environment 221:198–204.

Rusek, J. 1998. Biodiversity of Collembola and their functional role in the ecosystem. Biodiversity & Conservation 7:1207–19.

Ruthenberg, H. 1980. Farming Systems in the Tropics, 3rd edn. Clarendon Press. Oxford.

Sage, R.F. 2003. The evolution of C4 photosynthesis. New Phytologist 161(2):341–70.

Sánchez-Bayo, F. and Wyckhuys, K.A. 2019. Worldwide decline of the entomofauna: a review of its drivers. Biological Conservation 232:8–27.

Saxby, H., Gkartzios, M., and Scott, K. 2018. 'Farming on the edge': wellbeing and participation in agri-environmental schemes. Sociologia Ruralis 58(2):392–411.

Schader, C., Muller, A., Scialabba Nel, H., Hecht, J., Isensee, A., Erb, K.H., Smith, P., Makkar, H. P., Klocke, P., Leiber, F., Schwegler, P., Stolze, M., and Niggli, U. 2015. Impacts of feeding less food-competing feedstuffs to livestock on global food system sustainability. Journal of the Royal Society Interface 12(113):20150891.

Schaetzl, R.J. and Thompson, M.L. 2015. Soils: Genesis and Geomorphology, 2nd edn. Cambridge University Press, Cambridge.

Schäfer, R.B., van den Brink, P.J., and Liess, M. 2011. Impacts of pesticides on freshwater ecosystems. Ecological Impacts of Toxic Chemicals 2011:111–37.

Schiermeier, Q. 2009. Satellite data show Indian water stocks shrinking. Nature 460:789.

Schneider, R.W., Hollier, C.A., Whitam, H.K., Palm, M.E., McKemy, J.M., Hernandcz, J.R., Levy, L., and DeVries-Paterson, R. 2005. First report of soybean rust caused by *Phakopsora pach yrhizi* in the continental United States. Plant Disease 89(7):774–4.

Schofield, R. and Kirkby, M. 2003. Application of salinization indicators and initial development of potential global soil salinization scenario under climatic change. Global Biogeochemical Cycles 17(3).

Schütte, G., Eckerstorfer, M., Rastelli, V., Reichenbecher, W., Restrepo-Vassalli, S., Ruohonen-Lehto, M., Saucy, A.G.W., and Mertens, M. 2017. Herbicide resistance and biodiversity: agronomic and environmental aspects of genetically modified herbicide-resistant plants. Environmental Sciences Europe 29(1):5.

SedNet 2004. Report to SedNet: Contaminated Sediments in European River Basins. SedNet, Utrecht.

Semenov, M.A. and Shewry, P.R. 2011. Modelling predicts that heat stress, not drought, will increase vulnerability of wheat in Europe. Scientific Reports 1:66.

Senapati, N., Stratonovitch, P., Paul, M.J., and Semenov, M.A. 2018. Drought tolerance during reproductive development is important for increasing wheat yield potential under climate change in Europe. Journal of Experimental Botany 70(9):2549–60.

Seré, C., Steinfeld, H., and Groenewold, J. 1996. World Livestock Production Systems. FAO, Rome.

Shapiro, J. 2001. Mao's War Against Nature: Politics and the Environment in Revolutionary China. Cambridge University Press, Cambridge.

Shennan, C. 2008. Biotic interactions, ecological knowledge and agriculture. Philosophical Transactions of the Royal Society of London B: Biological Sciences 363(1492):717–39.

Shepherd, M., Pearce, B., Cormack, B., Philipps, L., Cuttle, S., Bhogal, A., Costigan, P., and Unwin, R. 2003. An Assessment of the Environmental Impacts of Organic Farming. A Review for DEFRA-Funded Project OF0405. ADAS Consulting, Mansfield, UK.

Shevade, V.S. and Loboda, T.V. 2019. Oil palm plantations in Peninsular Malaysia: determinants and constraints on expansion. PLoS One 14:0210628.

Schlink, A.C., Nguyen, M.L., and Viljoen, G.J. 2010. Water requirements for livestock production: a global perspective. Revue Scientifique et Technique 29:603–19.

Shukla, A. 2018. Evaluation of hexythiazox 5.45 EC against red spider mites (*Tetranychus urticae*) on okra. Journal of Entomology and Zoology Studies 6(2):43–6.

Shuster, W. D., Bonta, J., Thurston, H., Warnemuende, E., and Smith, D.R. 2005. Impacts of impervious surface on watershed hydrology: a review. Urban Water Journal 2(4):263–75.

Siebert, S.F. 2002. From shade-to sun-grown perennial crops in Sulawesi, Indonesia: implications for biodiversity conservation and soil fertility. Biodiversity & Conservation 11(11):1889–902.

Siebert, S., Portmann, F.T., and Döll, P. 2010. Global patterns of cropland use intensity. Remote Sensing 2(7): 1625–43.

Siemann, E., Tilman, D., Haarstad, J., and Ritchie, M. 1998. Experimental tests of the dependence of arthropod diversity on plant diversity. The American Naturalist 152:738–50.

Simard, S.W., Beiler, K.J., Bingham, M.A., Deslippe, J.R., Philip, L.J., and Teste, F.P. 2012. Mycorrhizal networks: mechanisms, ecology and modelling. Fungal Biology Reviews 26:39–60.

Smil, V. 1999. The detonator of the population explosion. Nature 400:415.

Smith, B.M., Chakrabarti, P., Chatterjee, A., Chatterjee, S., Dey, U.K., Dicks, L.V., Giri, B., Laha, S., Majhi, R.K., and Basu, P. 2017. Collating and validating indigenous and local knowledge to apply multiple knowledge systems to an environmental challenge: a case-study of pollinators in India. Biological Conservation 211:20–8.

Smith, B.M., Gathorne-Hardy, A., Chatterjee, S., and Basu, P. 2018. The last mile: using local knowledge to identify barriers to sustainable grain legume production. Frontiers in Ecology and Evolution 6:102.

Smith, P., Ashmore, M., Black, H., Burgess, P., Evans, C., Hails, R., Potts, S.G., Quine, T., and Thomson, A. 2011. Regulating services. In: UK National Ecosystem Assessment, chapter 14. UNEP–WCMC, Cambridge, UK.

Smith, R.G., Gross, K.L., and Robertson, G.P. 2008. Effects of crop diversity on agroecosystem function: crop yield response. Ecosystems 11:355–66.

Soane, B.D., Ball, B.C., Arvidsson, J., Basch, G., Moreno, F., and Roger-Estrade, J. 2012. No-till in northern, western and south-western Europe: a review of problems and opportunities for crop production and the environment. Soil and Tillage Research 118:66–87.

Socher, S.A., Prati, D., Boch, S., Müller, J., Baumbach, H., Gockel, S., Hemp, A., Schöning, I., Wells, K., Buscot, F., and Kalko, E.K. 2013. Interacting effects of fertilization, mowing and grazing on plant species diversity of 1500 grasslands in Germany differ between regions. Basic and Applied Ecology 14(2):126–36.

Soil Science Society of America. 2008. Glossary of Soil Science Terms 2008. ASA-CSSA-SSSA, Madison, WI.

Sölkner, J., Grausgruber, H., Okeyo, A.M., Ruckenbauer, P., and Wurzinger, M. 2007. Breeding objectives and the relative importance of traits in plant and animal breeding: a comparative review. Euphytica 161:273–82.

Springmann, M., Godfray, H.C.J., Rayner, M., and Scarborough, P. 2016a. Analysis and valuation of the health and climate change cobenefits of dietary change. Proceedings of the National Academy of Sciences 113(15):4146–51.

Springmann, M., Mason-D'Croz, D., Robinson, S., Garnett, T., Godfray, H.C.J., Gollin, D., Rayner, M., Ballon, P., and Scarborough, P. 2016b. Global and regional health effects of future food production under climate change: a modelling study. The Lancet 387(10031):1937–46.

Stanley, D.A., Gunning, D., and Stout, J.C. 2013. Pollinators and pollination of oilseed rape crops (*Brassica napus* L.) in Ireland: ecological and economic incentives for pollinator conservation. Journal of Insect Conservation 17:1181–9.

Stanley, P. 1995. Robert Bakewell and the Longhorn Breed of Cattle. Farming Press, Ipswich.

Staver, C., Guharay, F., Monterroso, D., and Muschler, R.G. 2001. Designing pest-suppressive multistrata perennial crop systems: shade-grown coffee in Central America. Agroforestry Systems 53(2):151–70.

Steffen, W., Richardson, K., Rockström, J., Cornell, S.E., Fetzer, I., Bennett, E.M., Biggs, R., Carpenter, S.R., De Vries, W., and de Wit, C.A. 2015. Planetary boundaries: guiding human development on a changing planet. Science 347:1259855.

Steffens, D., Hutsch, B.W., Eschholz, T., Losak, T., and Schubert, S. 2005. Water logging may inhibit plant growth primarily by nutrient deficiency rather than nutrient toxicity. Plant Soil and Environment 51(12):545.

Steinfeld, H., Gerber, P.J., Wassenaar, T., Castel, V., Rosales, M., and de Haan, C. 2006. Livestock's Long Shadow: Environmental Issues and Options. FAO, Rome.

Still, C.J., Berry, J.A., Collatz, G.J., and DeFries, R.S. 2003. Global distribution of C3 and C4 vegetation: carbon cycle implications. Global Biogeochemical Cycles 17(1):6–11.

Stoate, C., Henderson, I.G., and Parish, D.M. 2004. Development of an agri-environment scheme option: seed-bearing crops for farmland birds. Ibis 146:203–9.

Stuart, R.J., El-Borai, F.E., and Duncan, L.W. 2008. From augmentation to conservation of entomopathogenic nematodes: trophic cascades, habitat manipulation and enhanced biological control of *Diaprepes abbreviatus* root weevils in Florida citrus groves. Journal of Nematology 40(2):73.

Sullivan, W.M., Jiang, Z., and Hull, R.J. 2000. Root morphology and its relationship with nitrate uptake in Kentucky bluegrass. Crop Science 40(3):765–72.

Sutcliffe, O.L. and Kay, Q.O. 2000. Changes in the arable flora of central southern England since the 1960s. Biological Conservation 93:1–8.

Tabashnik, B.E., Carrière, Y., Soberón, M., Gao, A., and Bravo, A. 2015. Successes and failures of transgenic Bt crops: global patterns of field-evolved resistance. In: Bt Resistance: Characterization and Strategies for GM Crops Producing *Bacillus thuringiensis* Toxins, eds M. Soberón, Y. Gao, and A. Bravo, chapter 1. CABI, Wallingford, UK.

Tammeorg, P., Bastos, A.C., Jeffery, S., Rees, F., Kern, J., Graber, E.R., Ventura, M., Kibblewhite, M., Amaro, A., Budai, A., Cordovil, C.M.d.S., Domene, X., Gardi, C., Gascó, G., Horák, J., Kammann, C., Kondrlova, E., Laird, D., Loureiro, S., Martins, M.A.S., Panzacchi, P., Prasad, M., Prodana, M., Puga, A.P., Ruysschaert, G., Sas-Paszt, L., Silva, F.C., Teixeira, W.G., Tonon, G., Delle Vedove, G., Zavalloni, C., Glaser, B., and Verheijen, F.G.A. 2016. Biochars in soils: towards the required level of scientific understanding. Journal of Environmental Engineering and Landscape Management 25(2):1–16.

Tao, J. and Li, Y.O. 2018. Edible insects as a means to address global malnutrition and food insecurity issues. Food Quality and Safety 2:17–26.

Thanner, S., Drissner, D., and Walsh, F. 2016. Antimicrobial resistance in agriculture. MBio 7(2):e02227-15.

Thomasson, J.A., Shi, Y., Olsenholler, J., Valasek, J., Murray, S.C., Bishop, M.P., Blackmore, S., Gouache, D., Beauchêne, K., Mini, A., and Fournier, A. 2017. Autonomous air and ground sensing systems for agricultural optimization and phenotyping II. In: Proceedings Volume 10218. SPIE, Washington, DC.

Thorlakson, T. and Neufeldt, H. 2012. Reducing subsistence farmers' vulnerability to climate change: evaluating the potential contributions of agroforestry in western Kenya. Agriculture & Food Security 1:15.

Tilman, D. and Clark, M. 2014. Global diets link environmental sustainability and human health. Nature 515:518.

Tilman, D., Wedin, D., and Knops, J. 1996. Productivity and sustainability influenced by biodiversity in grassland ecosystems. Nature 379(6567):718.

Tilman, D., Fargione, J., Wolff, B., D'Antonio, C., Dobson, A., Howarth, R., Schindler, D., Schlesinger, W.H., Simberloff, D., and Swackhamer, D. 2001. Forecasting agriculturally driven global environmental change. Science 292(5515):281–4.

Tilman, D., Socolow, R., Foley, J.A., Hill, J., Larson, E., Lynd, L., and Williams, R. 2009. Beneficial biofuels—the food, energy, and environment trilemma. Science 325:270–1.

Tilman, D., Balzer, C., Hill, J., and Befort, B.L. 2011. Global food demand and the sustainable intensification of agriculture. Proceedings of the National Academy of Sciences 108(50):20260–4.

Tilman, D., Reich, P.B., and Isbell, F. 2012. Biodiversity impacts ecosystem productivity as much as resources, disturbance, or herbivory. Proceedings of the National Academy of Sciences 109:10394–7.

Tilman, D., Clark, M., Williams, D.R., Kimmel, K., Polasky, S., and Packer, C. 2017. Future threats to biodiversity and pathways to their prevention. Nature 546(7656):73.

Tooley, J.A. and Brust, G.E. 2002. Weed Seed Predation by Carabid Beetles. In: The Agroecology of Carabid Beetles, ed. J.M. Holland, 215–29. Intercept, Andover.

Tscharntke, T., Bommarco, R., Clough, Y., Crist, T.O., Kleijn, D., Rand, T.A., Tylianakis, J.M., van Nouhuys, S., and Vidal, S. 2008. Reprint of "Conservation biological control and enemy diversity on a landscape scale" [Biol. Control 43 (2007):294–309]. Biological Control 45(2):238–53.

Tuomisto, H.L. and Teixeira de Mattos, M.J. 2011. Environmental impacts of cultured meat production. Environmental Science & Technology 45(14):6117–23.

Turner, E.C., Snaddon, J.L., Fayle, T.M., and Foster, W.A. 2008. Oil palm research in context: identifying the need for biodiversity assessment. PLoS One 3:1572.

United Nations, Department of Economic and Social Affairs, Population Division 2017. World Population Prospects: The 2017 Revision. Key Findings and Advance Tables ESA/PWP/248. https://www.un.org/development/desa/publications/world-population-prospects-the-2017-revision.html.

Usher, M.B., Field, J.P., and Bedford, S.E. 1993. Biogeography and diversity of ground dwelling arthropods in farm woodlands. Biodiversity Letters 1(2):54–62.

van der Hoek, K.W. (1998): Nitrogen efficiency in global animal production. Environmental Pollution 102:127–32.

van der Putten, W.H., Bardgett, R.D., De Ruiter, P., Hol, W., Meyer, K., Bezemer, T., Bradford, M., Christensen, S., Eppinga, M., and Fukami, T. 2009. Empirical and theoretical challenges in aboveground–belowground ecology. Oecologia 161:1–14.

van Groenigen, J.W., Lubbers, I.M., Vos, H.M., Brown, G.G., De Deyn, G.B., and Van Groenigen, K.J. 2014. Earthworms increase plant production: a meta-analysis. Scientific Reports 4:6365.

van Huis, A. and Oonincx, D.G. 2017. The environmental sustainability of insects as food and feed. A review. Agronomy for Sustainable Development 37(5):43.

van Huis, A., Van Itterbeeck, J., Klunder, H., Mertens, E., Halloran, A., Muir, G., and Vantomme, P. 2013. Edible Insects: Future Prospects for Food and Feed Security (No. 171). FAO, Rome.

van Wart, J., van Bussel, L.G., Wolf, J., Licker, R., Grassini, P., Nelson, A., Boogaard, H., Gerber, J., Mueller, N.D., and Claessens, L. 2013. Use of agro-climatic zones to upscale simulated crop yield potential. Field Crops Research 143:44–55.

Vasey, D.L. 1992. Ecological History of Agriculture, 10,000 BC–AD 10,000. Purdue University Press, West Lafayette, IN.

Verheijen, F.G., Jones, R.J., Rickson, R.J., and Smith, C.J. 2009. Tolerable versus actual soil erosion rates in Europe. Earth Science Reviews 94:23–38.

Verhoeven, J.T. and Setter, T.L. 2009. Agricultural use of wetlands: opportunities and limitations. Annals of Botany 105:155–63.

Vermeulen, S.J., Campbell, B.M., and Ingram, J.S. 2012. Climate change and food systems. Annual Review of Environment and Resources 37:195–222.

Vetter, S.H., Sapkota, T.B., Hillier, J., Stirling, C.M., Macdiarmid, J.I., Aleksandrowicz, L., Green, R., Joy, E.J., Dangour, A.D., and Smith, P. 2017. Greenhouse gas emissions from agricultural food production to supply Indian diets: implications for climate change mitigation. Agriculture, Ecosystems & Environment 237:234–41.

Vickery, J.A., Feber, R.E., and Fuller, R.J. 2009. Arable field margins managed for biodiversity conservation: a review of food resource provision for farmland birds. Agriculture, Ecosystems & Environment 133(1–2):1–13.

Von der Ohe, P.C., Dulio, V., Slobodnik, J., De Deckere, E., Kühne, R., Ebert, R.U., Ginebreda, A., De Cooman, W., Schüürmann, G., and Brack, W. 2011. A new risk assessment approach for the prioritization of 500 classical and emerging organic microcontaminants as potential river basin specific pollutants under the European Water Framework Directive. Science of the Total Environment 409:2064–77.

Walker, B., Holling, C.S., Carpenter, S., and Kinzig, A. 2004. Resilience, adaptability and transformability in social–ecological systems. Ecology and Society 9:2.

Wambugu, F. and Wafula, J. 2000. Advances in maize streak virus disease research in eastern and southern Africa. Workshop Report, 15–17 September 1999, KARI and ISAAA Africa Center, ISAAA Brief No. 16. International Service for the Acquisition of Agri-Biotech Applications, Ithaca, NY.

Wang, L., Cai, Y., and Fang, L. 2009. Pollution in Taihu Lake China: causal chain and policy options analyses. Frontiers of Earth Science in China 3(4):437.

Wani, S.H., Sah, S.K., Sági, L., and Solymosi, K. 2015. Transplastomic plants for innovations in agriculture. A review. Agronomy for Sustainable Development 35(4):1391–430.

Watson, J.H.., Venter, O., Lee, J., Jones, K.R., Robinson, J.G., Possigham, H.P., and Allan, J.R. 2018. Protect the last of the wild. Nature 563:27–30.

Weltin, M., Zasada, I., Franke, C., Piorr, A., Raggi, M., and Viaggi, D. 2017. Analysing behavioural differences of farm households: an example of income diversification strategies based on European farm survey data. Land Use Policy 62:172–84.

Westerman, P.R., Wes, J.S., Kropff, M.J., and Van der Werf, W. 2003. Annual losses of weed seeds due to predation in organic cereal fields. Journal of Applied Ecology 40:824–36.

Weston, L.A. 1996. Utilization of allelopathy for weed management in agroecosystems. Agronomy Journal 88(6):860–6.

Westphal, C., Vidal, S., Horgan, F.G., Gurr, G.M., Escalada, M., Van Chien, H., Tscharntke, T., Heong, K.L., and Settele, J. 2015. Promoting multiple ecosystem services with flower strips and participatory approaches in rice production landscapes. Basic and Applied Ecology 16(8):681–9.

Wezel, A., Soboksa, G., McClelland, S., Delespesse, F., and Boissau, A. 2015. The blurred boundaries of ecological, sustainable, and agroecological intensification: a review. Agronomy for Sustainable Development 35(4):1283–95.

Whalon, M.E., Mota-Sanchez, D., and Hollingworth, R.M. 2008. Global Pesticide Resistance in Arthropods. CABI, Wallingford, UK.

White, P.J. and Brown, P.H. 2010. Plant nutrition for sustainable development and global health. Annals of Botany105(7):1073–80.

White, P.J., Bowen, H.C., Parmaguru, P., Fritz, M., Spracklen, W.P., Spiby, R.E., Meacham, M.C., Mead, A., Harriman, M., Trueman, L.J., and Smith, B.M. 2004. Interactions between selenium and sulphur nutrition in *Arabidopsis thaliana*. Journal of Experimental Botany 55:1927–37.

Whittlesey, D. 1936. Major agricultural regions of the Earth. Annals of the Association of American Geographers 26:199–240.

Wilkinson, T.J. 2013. Hydraulic landscapes and irrigation systems of Sumer. In: The Sumerian World, ed. H. Crawford, pp. 33–54. Routledge, Abingdon, UK.

Willer, H., Lernoud, J., Huber, B., and Sahota, A. 2018. The World of Organic Agriculture. Research Institute of Organic Agriculture (FiBL), Frick, and IFOAM–Organics International, Bonn.

Wilson, E.O. and MacArthur, R.H. 1967. The Theory of Island Biogeography. Princeton University Press, Princeton, NJ.

Wilson, M.C., Chen, X.Y., Corlett, R.T., Didham, R.K., Ding, P., Holt, R.D., Holyoak, M., Hu, G., Hughes, A.C., Jiang, L., and Laurance, W.F. 2016. Habitat fragmentation and biodiversity conservation: key findings and future challenges. Landscape Ecology 31(2):219–27.

Wilson, P. and King, M. 2003. Arable Plants: A Field Guide. Wild Guides. Princeton University Press, Princeton, NJ.

Wint, G.R.W., Rogers, D.J., and Robinson, T.P. 1997. Ecozones, farming systems and priority areas for tsetse control in East, West and Southern Africa. Consultancy Report by the Environmental Research Group Oxford Ltd and the Trypanosomiasis and Land Use in Africa Research Group, Department of Zoology, University of Oxford. FAO, Rome.

Wood, E., Harsant, A., Dallimer, M., Cronin de Chavez, A., McEachan, R., and Hassall, C. 2018. Not all green space is created equal: biodiversity predicts psychological restorative benefits from urban green space. Frontiers in Psychology 9:2320.

Wood, S.A., Karp, D.S., DeClerck, F., Kremen, C., Naeem, S., and Palm, C.A. 2015. Functional traits in agriculture: agrobiodiversity and ecosystem services. Trends in Ecology & Evolution 30(9):531–9.

Woodcock, B., Bullock, J., Shore, R., Heard, M., Pereira, M., Redhead, J., Ridding, L., Dean, H., Sleep, D., and Henrys, P. 2017. Country-specific effects of neonicotinoid pesticides on honey bees and wild bees. Science 356:1393–5.

Xiaoming, C., Ying, F., Hong, Z., and Zhiyong, C. 2010. Review of the nutritive value of edible insects. In: Forest Insects as Food: Humans Bite Back. Proceedings of a Workshop on Asia-Pacific Resources and their Potential for Development, eds P.B. Durst, D.V. Johnson, R.L. Leslie, and K. Shono, pp. 85–92. FAO Regional Office for Asia and the Pacific, Bangkok.

Yamauchi, T., Yoshioka, M., Fukazawa, A., Mori, H., Nishizawa, N.K., Tsutsumi, N., Yoshioka, H., and Nakazono, M. 2017. An NADPH OXIDASE RBOH functions in rice roots during lysigenous aerenchyma formation under oxygen-deficient conditions. Plant Cell 29:775–90.

Yeates, G.W., Bongers, T.d., De Goede, R., Freckman, D., and Georgieva, S. 1993. Feeding habits in soil nematode families and genera—an outline for soil ecologists. Journal of Nematology 25:315.

Zavaleta, E.S., Pasari, J.R., Hulvey, K.B., and Tilman, G.D. 2010. Sustaining multiple ecosystem functions in grassland communities requires higher biodiversity. Proceedings of the National Academy of Sciences 107(4):1443–6.

Zhang, L.Z., Van der Werf, W., Zhang, S.P., Li, B., and Spiertz, J.H.J. 2007a. Growth, yield and quality of wheat and cotton in relay strip intercropping systems. Field Crops Research 103(3):178–88.

Zhang, W., Ricketts, T.H., Kremen, C., Carney, K., and Swinton, S.M. 2007b. Ecosystem services and dis-services to agriculture. Ecological Economics 64(2):253–60.

Zhao, C., Liu, B., Piao, S., Wang, X., Lobell, D.B., Huang, Y., Huang, M., Yao, Y., Bassu, S., Ciais, P., and Durand, J.L. 2017. Temperature increase reduces global yields of major crops in four independent estimates. Proceedings of the National Academy of Sciences 114(35):9326–31.

Zoomers, A. 2010. Globalisation and the foreignisation of space: seven processes driving the current global land grab. Journal of Peasant Studies 37(2):429–47.

Zörb, C., Senbayram, M., and Peiter, E. 2014. Potassium in agriculture—status and perspectives. Journal of Plant Physiology 171:656–69.

Zougmore, R., Kambou, F., Ouattara, K., and Guillobez, S. 2000. Sorghum–cowpea intercropping: an effective technique against runoff and soil erosion in the Sahel (Saria, Burkina Faso). Arid Soil Research and Rehabilitation 14:329–42.

Zhu, Y., Wang, Y., Chen, H., and Lu, B.R. 2003. Conserving traditional rice varieties through management for crop diversity. Bioscience 53(2):58–162.

Index

Boxes, figures and tables are indicated by an italic *b*, *f*, or *t* following the page number.